ƒP

The
Modern Dog

A Joyful Exploration of How We Live with Dogs Today

STANLEY COREN

Free Press

New York London Toronto Sydney

FREE PRESS

A Division of Simon & Schuster, Inc.
1230 Avenue of the Americas
New York, NY 10020

First Free Press hardcover edition December 2008

FREE PRESS and colophon are trademarks of Simon & Schuster, Inc.

For information about special discounts for bulk purchases,
please contact Simon & Schuster Special Sales at
1-800-456-6798 or business@simonandschuster.com

Manufactured in the United States of America

10 9 8 7 6 5 4 3 2 1

Library of Congress Cataloging-in-Publication Data
Coren, Stanley.
The modern dog: a joyful exploration of how we live
with dogs today / Stanley Coren.
p. cm.
Includes bibliographical references and index.
1. Dogs. 2. Dogs—Behavior. 3. Dogs—Social aspects.
4. Human-animal relationships. I. Title.
SF426.C68 2008
636.7—dc22   2008034009
ISBN-13: 978-1-4165-9368-3
ISBN-10:   1-4165-9368-3

This book is dedicated to my niece and nephews,
Chad, Charna, Josh, and Michael

Contents

Preface

DOGS ARE INVENTED CREATURES—invented by humans in the sense that we have been continually shaping and changing them for at least 14,000 years. We are also continually shaping and changing the nature of our relationships to our dogs. We are always finding different ways to fit them into our lives and are also finding new jobs for them to do. This means that the modern dog, his world, and his involvements with humans are quite different today from what might have existed a century ago.

This is a book about the modern dog. It is meant to be a series of "snapshots" of the various ways that we interact with dogs, how society responds to dogs, how our relationships with dogs have changed over history, and where dogs fit into our personal and emotional lives. Dogs exist in our human world, which means that the only aspects of canine behaviors which are really important to the average person are those that affect the way that dogs and people interact. Often it is how people respond to and interpret the actions of dogs (and dog owners) that has a greater effect on a dog's life than the behavior patterns that have been programmed in the dog's genes.

Many sources of information tell us about the nature of human relationships with dogs. Obviously science has provided a lot of insight over the past few decades; however, folklore, religious writing, tradition, and even the actions of political bodies can all shed light on the dynamic interactions between humans and canines. Nonetheless, the exploration of the nature of the modern dog in this book has been designed to be much more of a romp than a formal exposition involving these sources of information.

Our look at the interactions of people and dogs will cover a broad range of topics. Here you will find the story of how certain types of dogs came to be, how dogs have become entangled in political and legal systems, and even how dogs may have influenced human evolution. Several chapters deal with some of the odd, compassionate, and even heroic behaviors dogs have been known to demonstrate. You will also meet a large collection of modern dogs and historic canines, including dogs that work, dogs that love, dogs that act reprehensibly, and dogs that will make you laugh. Alongside them you will see people who love, hate, work with, care for, and even obsess over dogs.

Since this is a book about how dogs fit into our society and culture as well as where they fit into our personal and psychological lives, it involves a lot of characters. There are some famous dogs, such as Strelka and Belka, the first living beings from Earth to go into orbit and survive, also dogs whose faithfulness or fighting spirit inspired statues to be erected in their honor, and you will hear about the real-life dog that was the basis for the much-loved story *Lassie Come Home*. Here you will also find heroic dogs, not so well known, but who have saved or protected human lives or more subtly mended the minds of people under stress. Along with the dogs come an array of humans with whom the dogs relate. They include kings and queens (Elizabeth I, Victoria, and Henry VIII), emperors (Frederick the Great, Napoleon, and Ming Ti of China), presidents and prime ministers (Winston Churchill, John F. Kennedy, both Roosevelts) and even a few gods, saints, and prophets (Anubis, St. Hubert, Buddha). There is also a collection of other interesting people who have had notable or unusual relationships with their dogs.

They include scientists, generals, physicians, schoolteachers, children, revolutionaries, and others.

Although this book will provide some important information about dog behavior, it is really designed to explore our relationships with and our emotional bonds to our dogs. Along the way you will learn how dogs can improve your physical and psychological health and that of your children. You will also learn how your dog can affect the way that other people view you or judge your place in society.

Each chapter of this book is meant to stand alone; you can read them in any order since no chapter depends on what was previously covered. In keeping with the lighter tone that I wanted for this book, I illustrated it in a range of styles, from some that appear to be woodcuts or engravings to some more modern pencil or pen and ink. I tried to make the style of the pictures fit the words or mood the story conveys. The pictures also allow you to browse through the book; using them as a guide, you can decide which chapter fits your mood or interests at that particular moment and simply start reading there.

There is a bit of personal history associated with this book. Early in 2002 I had a conversation with Connie Wilson, a beautiful blonde woman with a lot of drive and intelligence. Over the telephone she told me that she was going to try to start publishing a magazine called *Modern Dog* that would involve lifestyles and, of course, dogs. I laughed and told her that I had always wanted to do a book with that title and that I intended it to be an exploration of the human-canine relationship. Connie wanted me to do some writing for the magazine. I turned down the offer of a regular column, opting to write regular articles instead, and I have had one in every issue since *Modern Dog* began publication. The magazine has gone on to be quite successful and is now internationally distributed. Connie's love of dogs is shown by the fact that, as the magazine has prospered, she has used her association with it to sponsor a number of dog-related events.

During the six years of my own association with *Modern Dog* magazine I got to try out a number of themes and topics involving the shared lives of humans and dogs. The advantage of writing regularly for a magazine is that an author gets lots of feedback from the readership, in letters and e-mails (and a few bizarre telephone calls). These let him know which topics really interest dog owners and dog lovers. About one third of the chapters in this book actually started out as article ideas for *Modern Dog*, although all have been re-edited, expanded, and updated to take into account scientific advances, new information uncovered about the issues, and—most importantly—the wants, needs, and desires of the many readers who took the time to correspond with me.

As always, a book involves many more individuals than the author, and each contributed in a variety of ways. I would like to thank Connie Wilson and Jennifer Nosek at *Modern Dog* magazine for their warm interactions with me over the years. I would also like to thank my longtime friend Peter Suedfeld, who inspired and challenged me to write the chapter "*Semper* Fido," as well as providing the title. As always, many thanks go to my wife Joan, who had to deal with my fussing, this time not only about the words I was writing, but also the drawings I was creating. I greatly appreciate that she has still not yet resorted to a shotgun or a divorce lawyer to silence me. Finally there are the three modern dogs piled up at my feet as I write. I doubt that Dancer, Darby, and Banshee would understand my thanks for their supportive companionship; however, I know that they would appreciate a dog cookie just about now, so I hope that you will forgive me for stopping at this point to give them one. . . .

How Dogs Fit into the World of People

The Modern Dog

TODAY'S NEWS IS filled with concerns about genetic engineering and transgenic plants and animals. The creation of new strains of plants has led to concerns over the safety of our foods. It has also raised fears about the effect that these new plants will have on the insects that we depend upon to pollinate plants and how they may affect the environment and wild animals that come in contact with them. The creation of new strains of animals and microorganisms has led to fears that human "tampering with creation" may upset the natural balance that exists among species and ultimately result in the devastation of the world. Even Charles, the Prince of Wales, has gotten into the debate, claiming that the creation of new species "takes us into areas that should be left to God. We should not be meddling with the building blocks of life in this way." Familiar environmental organizations such as Greenpeace, and newer groups specifically formed to target this issue, such as Doctors for the Protection of the Environment, have spread the message that genetic engineering is a potentially dangerous instrument. However, genetic engineering and the manipulation of species is not a new thing. We have evidence that our genetic "meddling" goes back to at least 14,000 years ago, to when human beings created the first deliberately genetically-engineered organism—the dog.

The real truth of the matter is that, while God created man, man created dogs. This was not done in high-level biocontainment labs, but by seat-of-the-pants selection of desirable traits—which we can call "applied genetics." Over the last 140 centuries, we have continually changed the genetic nature of dogs in order to develop new breeds with different looks, behaviors, and abilities.

Despite the fact that most people believe that dogs have been around, unchanged, for eons, the modern dog is vastly different from the earliest domesticated canines. In fact, the dogs of today are different from dogs that existed only a century ago, and they continue to change to meet our lifestyles and our needs even today. The evolution of the dog is far from over, since human beings have now adopted the role played by nature in shaping this species. It is the action of people, not the evolutionary process of natural selection, which is changing dogs, and causing them to evolve in new directions. Humans now genetically influence not only the size and shape of dogs, but also their behaviors.

Man's first genetic intervention with respect to dogs was probably quite casual and accidental. Wolves and jackals, or other wild canines that were modern dog's predecessors were attracted to human campsites simply because primitive man paid little mind to sanitary practices. Because bones, bits of skin and vegetables, and other scraps of leftover food were likely to be scattered around human campsites, the ancestors of today's dogs (being ever food-conscious) learned that by hanging around man's dwellings they could grab a quick bite to eat now and then—without all that exertion and danger involved in actual hunting. While primitive man may not have been very concerned with sanitation, rotting foodstuff does smell bad and also attracts insects that make humans uncomfortable. So it is likely that dogs were initially tolerated around the perimeter of the camps because they would dispose of the garbage and eliminate these nuisances. The animals that were the least frightened by the presence of humans got most of the food, and those that were not aggressive and allowed humans to approach them were tolerated, had even more access to food, and thrived.

Once the wild canines that would eventually become dogs were attracted to human settlements, our ancestors noticed an added benefit. Early humans lived in dangerous times with large animals around that viewed humans as potential prey. In addition there were often other bands of humans with hostile intentions. Dogs hovering near the village looked on that settlement as their territory, which meant that whenever a stranger or some wild beast approached, they would set up a loud clamor by barking. This noise would alert the residents in time to rally some form of defense if it were needed. Since the dogs were always vigilant, human guards did not need to be posted throughout the night, thus allowing for more rest and a better lifestyle. It takes only a short mental leap to get from the concept of dogs guarding the edge of the village to the idea of a watchdog for an individual's home. Soon the dog's bark would serve the benign purpose of alerting the family to the approach of visitors (a sort of canine doorbell) or warning of the approach of a potential thief (a canine burglar alarm). This alerting function was clearly one of the motivations for domesticating dogs in the first place.

Now here is where the first genetic engineering comes into play. Once dogs were domesticated enough that humans could handle them and control their breeding, we could start to tinker with and modify the species. Obviously, for personal and community security, the most effective dog is one with a loud, persistent bark. So early humans began a selective breeding program to create such dogs. A dog that barked loudly was kept, nurtured, and bred with others that also barked. One that did not bark was simply disposed of as useless. Thus "barking genes" were strengthened in dogs to the degree that today one of the distinctions between wild canines and domestic dogs is the fact that domestic dogs bark a lot, and wild canines seldom do.

Subsequent developments of dogs were much more conscious and deliberate. In some ways the genetically engineered dogs that later appeared often seemed to be more like "inventions" than domesticated animals. This is because we created or modified dogs not just to fit our immediate needs but also to fit our current technology. Perhaps the best example of how the changing nature of dogs and the changes in technology go hand in hand can be seen in the evolution of the gun dog.

The earliest hunting dogs were hounds. Hounds naturally fall into two clear groupings based on their behavior and other characteristics. The first group is the *scent hounds*—dogs that are supposed to track their quarry by the faint odor they leave as they move over the landscape. This group includes bloodhounds, foxhounds, black-and-tan coonhounds, bassets, and beagles. The second group, *sight hounds,* has keen eyesight and tremendous speed, and includes greyhounds, salukis, Irish wolfhounds, Scottish deerhounds, and Afghan hounds. Their task is to visually locate their quarry in the distance and run it down. Hounds were initially designed for a world with primitive technology, where the main weapons used in hunting were spears and bows; there were no firearms. As independent hunters, hounds found their quarry and dispatched their target when they caught it, without any human intervention or cooperation. In most cases, the hunter merely had to catch up with the hounds before they had eaten the game. The human hunter's skills were needed only if the game found refuge in a den, burrow, or tree or if the animal was too large and dangerous for the dogs to deal with on their own, resulting in the need for human weaponry.

The invention of firearms changed the kind of dogs that humans needed. The earliest gun dogs were the pointers, designed to work with muzzle-loading muskets. These guns were clumsy and difficult to use. In order to load them, you had to take out your powder horn and dump some gunpowder down the barrel. Then you had to wrap a lead ball with a bit of oiled paper or cloth, which also had to be placed in the gun barrel. Next, you had to take the tamping rod from its holder against the barrel and tap down the shot and powder. Then you had to put the ramrod and powder horn back into their places. Finally, you needed to cock the hammer, aim, and fire. This whole process of just getting the gun loaded and ready to shoot could take thirty seconds or longer. After all of this effort, you had a weapon with an accurate range of only about 25 to 50 yards.

The pointer was biologically engineered to accommodate this kind of technology. It is a dog designed to move slowly and silently. Pointers are supposed to find a bird, mark its location by staring directly at it, then to hold this pointing position for a reasonably long time. You must understand that the hunter had only one shot. If he missed, the reloading process would obviously take too long to permit a second chance. The best strategy was for the hunter to get as close as possible before flushing the bird. Then, after a lucky shot actually killed some game (because as a species we humans are so lazy), the pointer was also expected to go out, collect the bird, bring it back intact, and place it in his master's hand.

Over time, weapons technology improved and guns became more efficient. Most importantly, they were faster loading, and had better range and accuracy. To match this new equipment we developed a new kind of dog, the setter. Setters move along much more quickly than do the pointers, and you can tell how near

Slow-moving pointers were ideal for hunting with muzzle-loading weapons.

they feel they are to the game by the beat of their tails. Their tails wag faster and faster as they get closer to the game. If you have a pair of setters you can locate where the birds are hiding very precisely. Since both dogs will freeze on point with their heads looking in the direction of the bird, you can use their lines of sight to triangulate the position of your quarry. You can actually sight down their heads, and where the two lines of sight cross is where the dogs think the bird is located.

Technology continued to progress, and eventually bolt-action and semiautomatic guns that could quickly fire several rounds were invented. Shotguns were also improved to have a longer range and the ability to fire more than one shot. With this new equipment the exciting sport of hunting behind spaniels became possible. Spaniels are totally undisciplined hunters. They just quarter the ground in front of the hunter and flush anything that is there without any warning. You've got to have a good gun if you're going to be able to get any kind of hit with this kind of hunting.

One of the most recently bioengineered developments in gun

dogs is the retriever. The retriever was invented because more land was being cultivated and cities and towns were springing up as a consequence of the industrial revolution. That meant that there was not a lot of free-range land around to use for hunting. This forced the nature of modern hunting to change. More and more hunters began to hunt on scrap land that had no industrial or agricultural value. This is often the land on the margin of swamps and marshes. Hunting for birds in that environment involves the use of blinds set up near a body of water where the hunters would wait for the ducks instead of going to where they thought the birds might be. Under these circumstances you need a dog that will remain quiet and be totally attentive to the handler. The dog must be willing to sit patiently for long periods of time, never trying to flush nearby birds, never running around trying to point at things, nor barking in excitement. Any such activities would simply chase away any incoming waterfowl. The retriever is supposed to stay quiet until it is needed, and then it must be willing to go through swamps or swim through cold water to bring a bird back to its master. The fact that these dogs must wait for their human master's cue to act explains why retrievers are among the most obedient and attentive dogs in all of dogdom.

Canada is responsible for the newest genetic invention in the retriever group—the Nova Scotia duck tolling retriever. It is a handsome orange-and-white-colored dog that stands about 19 inches high and weighs about 45 pounds. Hunters sitting in blinds, or floating on boats waiting for waterfowl to fly within range, would often get impatient, especially when the weather was damp or cold. To attract birds and get them to fly within gunshot range, they tried using fancy decoys that they floated near their blinds, and these helped a bit. But at some point, a chance observation led to the concept of a tolling retriever. The word "toll" means to entice or attract, in much the same way that church bells toll to attract people to come to services.

Someone noticed that pairs of foxes had developed a method for luring ducks. One fox basically cavorts about on the shore, spinning and turning, with its white-tipped tail held high like a flag. All of this is done within clear sight of the ducks, which float far out in the water. Curious about all of this activity, ducks swim closer and, when they come close enough to the marsh grass and reeds near the shore, the fox's mate springs out of ambush to catch the nearest bird.

To imitate the fox's success, Eddie Kenny of Yarmouth County in Nova Scotia attempted to breed a dog that not only resembled the fox with its red-orange coloration and the white flash on the tip of its tail, but also mimicked the fox's behavior, with the dog dancing and twirling when it got excited. This tends to attract not only ducks floating nearby, but also those flying overhead. The curious ducks come closer to observe what is happening, and when they are in range the hunters can shoot them. Once the duck is downed, the little dog is expected to act like a traditional retriever and bring the bird back to the hunter. Eddie Kenny and several other breeders who helped invent the "Toller" were still alive in 1945 when the breed was finally registered by the Canadian Kennel Club.

Outmoded technologies such as seventy-eight RPM record players, rotary dial telephones, or pedal-driven sewing machines are often discarded. Similarly, dog breeds that are no longer suited for the current level of technological advancement simply disappear. Take, for example, the Spanish pointer. Once one of the most popular hunting dogs, perfect for the era of the muzzle-loading musket, it is often seen in early paintings of hunting scenes. Slow and quiet, it was the most meticulous of the pointers. Now, these pointers are effectively extinct, because they were just too slow for today's impatient hunters with their new and better hunting equipment. There are still pointers, but they are English pointers, or German shorthaired pointers which

The most recently invented hunting dog is the Nova Scotia duck tolling retriever.

move more quickly and are better adapted to the hunting weapons of today.

It sort of gets you to thinking, doesn't it? Weapons technology, including hunting guns, will continue to evolve. We can only speculate about what kind of gun dogs will be needed to achieve the full potential of hunting in the Star Wars era—when hunters can stand on the roofs of their tall apartment buildings while they try to bring down birds with their laser-phaser-pulse guns.

Perhaps our faith in biogenetic engineering would be improved if we recognized that, for those of us who don't hunt, some dogs have also been designed to fit the couch-potato mentality of our television-addicted era. It is a wonder to me that starting with the DNA of a wolf, we have spent 14,000 years of biotechnology and genetic manipulation specifically to create the broad variety of modern dogs, which includes the little chestnut-and-white Cavalier King Charles spaniel that is right now gently snoring with his head resting against my foot. He has no hunting or retrieving instincts in him at all, but he fits into my contemporary lifestyle by serving as an extremely effective stress reducer for me. Despite the warnings against genetic engineering, I am glad that we "meddled" with the building blocks of life to produce my little companion.

Chapter 2

Why Neanderthals
Don't Rule the World

SCIENTISTS WHO STUDY evolution are often puzzled by the fact that the scrawny, flat-faced primates, with little hair and weak jaws, that we call Homo sapiens are the dominant species on this planet. In the distant past, Homo sapiens' most significant competition was a race of more robust, powerful, hairier primates with strong jutting jaws. The strength and ruggedness of the Neanderthals certainly seemed to give them the early survival advantage. The surprising answer to this evolutionary conundrum may actually lie in our ancestors' relationship with the ancestors of our pet dogs.

Clearly that line of Homo sapiens, beginning with the Cro-

Magnons (named after the cave of Cro-Magnon in southwest France, where the first specimen was found), survived as our most immediate ancestors while the Neanderthals did not. If you ask people why that was the case, you will probably be told that it is because we Homo sapiens had a bigger brain, were smarter, were tool makers, had better weapons, and had an organized social system that the Neanderthals lacked. Unfortunately this belief probably comes from the film and TV stereotypes that depict Neanderthal cavemen as shambling, aggressive, and stupid. The truth of the matter is that Neanderthals may have been more advanced than Cro-Magnons in many ways. For example, they actually had larger brains than their Homo sapiens neighbors.

The Neanderthal's physical appearance tends to catch our attention first, however. Although somewhat shorter than modern humans, Neanderthals were quite powerful, with heavy bones and large muscles. Their skulls were marked by a thick protective ridge over the eyes and their powerful, muzzle-like jaws could serve as an additional weapon or tool. In that muzzle was a nose that was larger and contained many more scent receptors than we have today, giving the Neanderthal a sense of smell superior to that of modern man.

Because of their large brains, Neanderthals were quite clever and made effective tools and weapons, including stone and bone hammers; spears, some with sharpened stone points; and stone axes and knives. They also made useful domestic tools, like needles and awls, that allowed them to sew and piece together clothing as protection against the cold weather of their time. They lived in small communities that seem to have had an organized social life. They apparently also had religious beliefs and rituals, since Neanderthal graves have been found containing flowers, tools, and other personal possessions. It appears that the Neanderthals developed the concept of trade, since there is evidence that they traded goods with nearby Homo sapiens as well as other Neanderthal groups.

Neanderthals hunted in coordinated groups. Their spears and axes were effective on large animals, and we know that they attacked and killed creatures as big as the woolly mammoth. Hunting large animals was actually the most efficient means of staying alive, since one kill could provide food for many days or weeks in a small community. Stone axes and spears were not the right technology to hunt smaller prey, which moved too swiftly and were difficult to approach closely enough to be speared or hit with an ax.

It was the coming of the Ice Age that provided the real challenge. Ultimately, with bands of both Neanderthal and the Cro-Magnon hunters hunting big game animals and with the cooling climate reducing the vegetation that served as their food, the numbers of larger prey diminished sharply and many species became extinct. Survival of the Neanderthals and Homo sapiens then depended on their abilities to adapt to the new environment and to find food sources other than large animals.

By the time this crisis occurred, our ancestral Homo sapiens already had a population advantage over Neanderthals. The size of a population is a good indication of how successful any species is in an evolutionary sense. So how could the weaker, smaller-brained species have accomplished this? In order to answer this question we have to have a better understanding of how evolution works.

Most people believe that species evolve by developing characteristics that allow them to adapt better to their environment. If these useful characteristics can be passed on through their genes, it improves the survival of that species. However, there are two important components to consider. First, there is the individual's *genotype*, which is the inherited set of instructions coded in DNA. Second, there is its *phenotype*, made up of the actual characteristics and qualities of that animal that you observe. The phenotype includes an individual's shape and size, the course of its development, and the behaviors that it actually engages in. Think of it as the difference between the genes that are inherited by an individual and what that heredity actually produces in a particular individual living in a specific environment.

When animals live in close association with each other, their communities act as if each has a phenotype. Like an individual's phenotype, this community phenotype is the collection of all its attributes. Natural selection (the mechanism of evolution) actually works on phenotypes, since those are the characteristics that affect survival. Theoretically this means that when two species live together, there is the possibility that the community can evolve, and all of the species that live in that community can begin a process of coevolution.

Richard Dawkins, a British evolutionary biologist at Oxford University, has suggested that evolution can act on the extended phenotype of a community in order to preserve all of those behaviors of its members that increase the likelihood that the group will survive. Whether the members of the group happen to be of the same species or not doesn't matter. If the community's phenotype depends on behaviors and characteristics that some individual members can pass on genetically, then the community as a whole can evolve through natural selection in the same way an individual species can. Furthermore, this evolutionary process can occur without any deliberate actions or intentions of the community members. It is through just such a process that dogs played their role in human evolution.

Faced with disappearing large game, some groups of Cro-Magnons began to try several new tactics. They formed home camps or bases that were relatively fixed and permanent, where the individual band members could actively share in various tasks, such as the gathering of available food from local plants, an activity that eventually led to a primitive form of agriculture.

The establishment of such a fixed residential area led to the growth of garbage dumps around the outer limits of the village, which naturally led to an infestation of opportunistic scavengers. While these included mice and rats, they also included wolves and jackals, the ancient precursors of dogs.

Similar situations occur even in our modern society. Researchers have shown that in Italy the garbage dump serves as the primary food source for local wolves and unowned dogs. The villagers are usually unaware of this circumstance since the animals feed at night. Biologist Raymond Coppinger tells a story of a researcher turning on a spotlight at a local garbage dump at two in the morning and catching a wolf with his mouth full of cast-off spaghetti.

The canines that surrounded these early settlements became part of the loose community of Homo sapiens. These grandfathers of modern dogs had a much better sense of smell than Homo sapiens (even better than their Neanderthal competitors), and a better sense of hearing. This turned these scavengers into an effective canine alarm system since they could detect the approach of hostile people or animals even better than human sentinels might. Their barking provided a timely warning, allowing villagers to survive raids by competitors who did not have dogs. If we believe the concept of an extended phenotype, the resident dogs and humans were now part of a community that was evolving. In essence, the dogs and humans entered into an evolutionary contract. The human beings in dog-infested settlements had a higher survival rate because the dogs used their better sensory perception to serve as sentinels. Human beings could concentrate on organizing their home base to survive attacks and create weapons and tools. The humans did not have to be eternally vigilant, which meant that they also got more useful sleep, which improved their health and functional intellectual abilities. This more efficient home base in turn provided more reliable food sources for scavengers, including the evolving domestic dog.

Dogs surrounding the settlement would raise the alarm if anyone approached.

Whenever humans rallied to the sound of a canine alarm, indicating the presence of a predator, they not only protected their fellow human inhabitants but also protected their companion dogs as well. The dogs improved their own survival rate even more by developing cooperative traits that made them indispensable and endearing to their early human associates.

The fossil record shows that around 14,000 years ago, the shape and size of the wolf skulls found around ancient human settlements changed, indicating that they had become domestic dogs. No evidence of domesticated dogs appears around any Neanderthal sites.

At about this same time, an exciting new development in human weaponry occurred, namely the appearance of small, long-distance projectiles (arrows) with sharp stone tips. While this provided a major safety advantage by allowing hunting from a distance, there was an even more important use for arrows. Most of the really big game had disappeared, but a swift-moving arrow shot from many feet away permitted the successful hunting of smaller prey.

The problem with arrows is that they wound, and only rarely kill outright. Such a wound makes it possible for the hunter to get close enough to the injured animal to finish the job. Unfortunately, wounded animals can get away, and can put a long distance between themselves and their pursuers. This meant that injured game often had to be pursued and tracked, sometimes for quite a way, before it could be cornered or immobilized so that the final strike could be delivered. Furthermore, an animal's trail is easily lost in dense brush or over rocky regions. Fortunately dogs are the masters of tracking wounded animals. Some scientists believe that, without the assistance of dogs, bow-and-arrow hunting never would have been successful.

The benefits provided by watchdogs and hunting dogs could well have been sufficient to allow early humans enough of a survival advantage to outlast and replace the dogless Neanderthals. But humans living with dogs may have had another subtle but more important evolutionary advantage. David Paxton of the Australian National University has offered a speculative, controversial theory of what might have been the ultimate evolutionary development that guaranteed the survival of Homo sapiens as the dominant species on Earth.

Paxton begins with the idea that humans and dogs co-evolved. Because humans had smaller heads than Neanderthals, they did not require thick neck muscles for support, which allowed space for the development of vocal cords and other structures to make precise sounds. Since the jaw was no longer a tool and a weapon, it was possible to develop more flexible lips and face muscles, a combination that allowed for increased communication ability through a greater variety of facial expressions, and most importantly, through the conscious control of sounds that ultimately led to the development of language. The appearance of more complex communication abilities allowed for more helpful social interactions, more coordination in group activities, and the transmission of useful information which, together, would have greatly improved human survival.

If the appearance of better communication abilities is part of the co-evolution of humans and dogs, as Paxton suggests, then dogs should also have evolved in a manner to take advantage of human communication. Brian Hare at Harvard University has been looking at the communication and cognitive skills of various types of animals and has found that domesticated dogs are superior to all other animals at picking up on subtle, nonverbal human communication cues, such as pointing or facial expressions. These abilities appear in puppies, whether they have been socialized to be around humans or not. In comparison, wolves show no such talent, even if they have been raised with humans all their lives. This suggests that responding to human communication is a hereditary trait of domesticated dogs that may well have come about as dogs and humans coevolved.

If we take all of these influences into account, it is easy to put together a possible scenario to explain how the community of humans and canines won the planet from the Neanderthals. Humans would have been healthier because of more effective hunting with dogs. Well fed, they would have been less susceptible to disease, and their settlements would have supported larger populations. Having lost the large game on which they depended, Neanderthals might have been driven to raid human communities for supplies, but their attacks would not have been successful because the human communities' dogs would have had early warning of the attacks, which would have enabled Homo sapiens to rally an effective defense. After a few such raids Homo sapiens would doubtless have come to view Neanderthals as mortal enemies. With the humans' improved communication abilities, they could have coordinated larger groups to counter-attack their neighbors who were weakened from hunger. The absence of dogs around Neanderthal camps would have meant no ring of sentinels, and any human raids could have surprised them. Such a pattern, over time, could easily have led to the extinction of the Neanderthals.

If this scientific speculation is correct, then, while it may appear that humans have controlled the dog's evolution, dogs may well have played a major role in determining the evolutionary path of humans. In essence, dogs decided which of the species of early humanoids would control this world, and the Neanderthals lost out.

Chapter 3

The Children of Anubis

SOME 4000 YEARS ago, the Pharaoh of Egypt, Antefa II, had a dog named Tekar. The dog slept on his bed and was emotionally dear enough to the pharaoh that, in one portrait of the king, Tekar was drawn standing between his legs.

For most early civilizations we know little about the intimate interactions between prominent people and their dogs, and even less about the relationships of ordinary people and their pets. Most early histories tended to focus on rulers and other high-ranking figures, and dealt only with major events, such as wars. However, ancient Egypt is different. This civilization left us a great deal of information in the form of texts, tomb drawings, and artifacts that allow us to piece together a picture of a society in which the human-canine bond was strong, and dogs were kept, loved, and even honored by people of all classes.

The first Egyptian dogs of which we have a record appear in hunting scenes drawn around 4240 B.C. Egyptians who could afford it loved to hunt gazelle, ibex, antelope, and other fleet desert animals, which required fast-running dogs with large lungs, narrow waists, and long legs. These were the forerunners of our sighthounds—dogs that use their eyes to detect their prey, and their speed to capture it. The greyhound is a direct descendant of those dogs, with only a few slight differences in characteristics. Our modern greyhound has floppy ears, while in ancient Egypt it could have either large pricked ears or lopped ears. In addition, the faces of these dogs could be either long and narrow (very much like a jackal) or somewhat shorter and wider. One distinctive characteristic of all of these ancient dogs was that their tails were held high and often curled over their backs. The Pharaoh and Ibizan hounds of today look much like the ancient prick-eared, long-faced dogs, while greyhounds, whippets, and salukis look more like the lop-eared versions, with faces not quite as elongated and all of these carry their tails lower.

Over time the number of different breeds of dogs increased. For example, the Hyksos, invaders from the northeast, introduced large, square-jawed mastiff-type dogs, which could be used as weapons of war and also were effectively employed to hunt larger game, such as lions. In addition, representations of smaller dogs that seemed to have no definable function other than to serve as pets and companions began to appear in early drawings of Egyptian life. In 1836, Sir John Gardner Wilkinson cataloged the many dogs depicted in ancient Egyptian tombs, and his collection also included small whippet-like dogs, basenjis, something much like a bassett hound, and even a dog that somewhat resembles a dachshund.

You may have read that the ancient Egyptians were one of the few cultures to actually worship dogs as gods. This is because

Anubis, the god associated with resurrection, who guides the dead to the place where they will be judged, has the head of a dog. Some say that the head of Anubis is really a jackal; however, another god, Upuaut, truly has the head of a jackal, while the face of Anubis is really much like the prick-eared, long-faced greyhounds depicted on tombs. The truth of the matter is that the Egyptians did not have animal gods but believed that the spirit of each god resided in particular animals, and such animals were sacred to him. To attract the attention and favor of a particular god, the Egyptians offered sacrifices of that god's animal.

Anubis became the god who guided the dead, partly because of his part in an earlier legend in Egyptian mythology. Osiris, the sun god, was responsible for growing plants as well as judging the dead. He was married to Isis, the great mother goddess of the earth and moon. Osiris was murdered by Set (the brother of Osiris and the god of evil), who then threw his body into the Nile. Isis found the body and took it to the god of knowledge to restore him to life, but Set discovered them, stole the body, tore it into fourteen pieces, and scattered them all over the world. Isis set out to find all the parts of her husband's body and reassemble them so that he could be sewn together by Anubis, lord of the dead. It was believed that Anubis was the one who invented the process of mummification as part of his efforts to bring Osiris back to life. He embalmed the body of the god, swathed it in the linen cloths that had been woven by the twin goddesses, Isis and her sister Nephthys, and therefore ensured that the body would never decay or rot. After he brought Osiris back to life, Anubis ceded his throne to him in the underworld. In the temples that performed dramas reenacting the legends of Isis and Osiris, a priest impersonating Anubis would wear a mask with the face of a dog and, at the climax, reveal a little boy who had hidden beneath his robes to symbolize the rebirth of the lost god.

Sadly, the association of dogs with Anubis may have cost the lives of many thousands of canines. The city of Hardai became known as Cynopolis (City of Dogs) by the Greeks, because of the temples to Anubis there. Anubis was responsible for weighing a deceased person's heart (which supposedly contained all of the deeds the person had committed in life) and balancing it against the *feather of truth*. These results were then given to Osiris to determine the fate of that soul. Since Anubis calibrated the scale and gave the report, it was believed that he might be convinced to bias the results a little bit to benefit those that he favored. People looking to benefit from this flocked to Cynopolis to make offerings for themselves or for recently deceased loved ones and in the process they used sacrificial dogs. The temples of Anubis became overrun with dogs that wandered around looking for handouts and sadly became preferred targets for sacrifice since it was believed that they were especially holy. The temple dogs for these ritual gifts had to be mummified, but the natural death rate of temple dogs was not high enough to keep up with the demand for dog mummies. For this reason the temples expanded to include what amounted to factories that produced mummies of young dogs bred in temple kennels. The priests of Anubis created assembly lines made up of specialized staff who killed, gutted, prepared, dried, and applied the final decorative wrappings and ornamentation to the bodies of dogs so that worshippers would have a genuine temple-dog mummy to give as an offering. Later archaeologists turned up so many of these dog mummies around Cynopolis that they began to view them as a nuisance, rather than as significant findings filled with information about the beliefs, hopes, and fears of an ancient people. Because of this they allowed tons of animal mummies to be exported to England, where they were pulverized and sold as garden fertilizer.

In spite of its industrialized ritual killing of dogs, Egypt still provides the first historical records indicating an emotional bond between people and their dogs. By law, the killing of an owned dog, that is, one with a collar, was punishable by death. If a household dog died, grief was expressed as if a person had died: the dog's master shaved off his eyebrows and the hair on his head and body. (If a cat died, a man was required to shave off only one eyebrow.) Virtually every large ancient city had a cemetery where the mummified remains of well-loved dogs could be placed along with inscriptions containing words like "beloved," "companion," "blessed," "faithful," and "we will be united in paradise."

The love for a dog was expressed in many ways. Many pharaohs were buried with effigies of their dogs. Ramses II was buried with the names and statues of four of his dogs, and Tutankhamen with images of two of his dogs.

Perhaps the greatest expression of caring was to have the dog's body mummified and interred with its master, sometimes on a reed mat at the foot of the bed or near the feet of the master's sarcophagus. Cheops, the pharaoh who built the Great Pyramid at Giza, had a dog named Abarkaru, described as the pharaoh's companion and protector. Abarkaru's mummy was placed in the same chamber where his king lies.

In a few instances the mummy of the dog was actually placed in the sarcophagus. In one woman's burial, the container with the dog was laid by her feet and carried the inscription "a hound of the bed who his mistress loved." In another the mummy of a man rests next to his dog with the words "An unfailing friend. His name was Abu."

An Egyptian folktale, sometimes called "The Foredoomed Prince," demonstrates the value that was placed on the friendship of a dog in those ancient times. The story begins with the observation that Pharaoh's sister had been thought to be barren, but late in her life she had a son whom she named Baraka, meaning "gift" or "blessing." The priests were called in to bless the child and to foretell his future. The soothsayer among them declared, "The death of this child will be determined by the actions of a crocodile, or of a serpent, or of a scorpion, or of a dog."

To protect her child, the sister asked Pharaoh to provide her son with a large house near the edge of the royal quarter of the city. The boy would be raised in isolation and never allowed to wander in the desert, near the river, or in the streets where he might encounter the creatures that could mean his death. One day, Baraka stood on the roof terrace and looked at the nearby street where a cart was rolling by. In the cart were a young boy and girl, and they were playing with a litter of puppies; he could hear their laughter.

Baraka asked, "What are these things? Could I have one like that which might make me laugh?"

Pharaoh learned of this and consulted the priests, who told him that allowing the child to be near any of the four creatures that had been named might place the boy in immediate danger. On the other hand, they said, fate could not be cheated and if the child was doomed the god Set would claim him no matter what was done.

So the king chose a puppy sired by his favorite dog. He named him Uzat, after the eye of the god Horus who protects people from evil. The king then sent the dog to be a pet for his nephew, "lest his heart be sad."

Uzat became Baraka's companion and best friend. One day the boy managed to sneak out of the house, and he and his dog explored the nearby bank of the Nile River. As the sun rose toward noon, the boy sat in the shade and fell asleep. Suddenly he was awakened by Uzat's barking. When he stood up, he saw a crocodile that had reached the shore of the river and was charging toward him. The dog's barking not only alerted him, but distracted the crocodile long enough to give the boy a chance to run to safety, followed closely by his vigilant dog. Uzat was rewarded with a bowl of fine meat that night.

Baraka was alerted by Uzat's barking in time to escape the crocodile.

As was the custom of the time for members of the royal family, the prince's marriage was arranged by the pharaoh. When he first met Shepsit, his bride-to-be, they spoke of many things, including the fate that had been predicted for him.

Shepsit was horrified. "Then you must kill that dog to keep yourself safe," she said.

Baraka replied, "I will not kill my dog. I raised him from when he was small. He is my friend, and he saved my life from the crocodile."

"I do not want a husband who so carelessly flirts with death," she replied.

"Would you rather have a husband who cares nothing for his friends and would turn on those that he loves?"

Shepsit heard his words and relented so that they could marry. Uzat grew old and died and was replaced by one of his puppies, that they also called Uzat. When Baraka went off to serve in the army, this Uzat came with him. One morning, as Baraka awakened and reached for his boots, Uzat suddenly began to growl. The young man wondered if this was the sign of the long-feared attack that would kill him. He reached for the boot again, and the dog charged across the room, grabbed the boot by its tip, and shook it from side to side. Out of the boot fell two scorpions that might well have killed Baraka had he stepped in the boot.

Years passed, and there was yet another Uzat sired by the previous dog living in Baraka's home. Shepsit and Baraka sometimes talked about the fate predicted by the priests, but he still insisted on having his dog with him. One day a servant brought in a basket of figs that had been harvested nearby. Baraka thought that they looked quite ripe and good and was reaching for one when Uzat leapt forward and knocked him down. As the dog regained his feet, he spun around and in the process the basket of figs was knocked over. One servant ran over to help his master up. Another went to retrieve the figs but gave a cry of fear when a deadly asp wriggled out of the basket. It was a serpent that could well have killed Baraka. Once again his life had been saved by his dog.

More years passed, and there was yet another Uzat sitting by Baraka's side. The prince was now growing old and ill. At the insistence of his wife they called in the healer and a priest. Baraka turned to the priest and asked, "My fate was to have my death determined by the actions of a crocodile, serpent, scorpion, or a dog. I am now old, and if the healer is correct, I will die of causes associated with advanced age. Does this mean that the original prediction of my death was wrong?"

The priest smiled and said quietly, "No, your fate was as it should be. Because you accepted the friendship of a dog, and were true and loyal to him, you were saved from an early death by a crocodile, a serpent, and a scorpion. Thus it is the actions of a dog that have, in fact, determined the nature of your death. The dog's actions are what have caused you to die of old age rather than some other cause. Through the friendship that you have shared with a dog, you have been blessed by Anubis with a long life, and most likely you will be rewarded for that loyalty in the afterlife."

Chapter 4

The Patron Saint of Dogs

ONE EVENING I attended a reception after the wedding of the daughter of one of my colleagues. He is a devout Catholic, and I found myself standing in a corner with three Catholic priests, one of whom was my friend's oldest son. He was now referred to as Father Rommy and was in the company of a very young priest, Father Allen, a family friend who had gone through school with the bride. The third, much older man was the parish priest, Father Francis, who had conducted the marriage ceremony.

While we stood and sipped punch I reminisced about a trip that I had taken through rural North Carolina. It had been early in November a number of years ago. Riding down a two-lane road I suddenly came upon a large crowd standing in front of a small church. Since I was alone and not on any schedule I stopped to see what was going on. The crowd consisted of all kinds of people and their dogs: children with pet dogs, shepherds with sheepdogs, hunters with hounds, even some people with large dogs that looked like guard dogs. When I asked what event had brought all of those canines to church, I was told that November 3 was the feast day of St. Hubert, the patron saint of dogs.

The son of the Belgian Duke of Guienne, Hubert was boisterous, self-indulgent, and, as a young man, someone who dearly loved to hunt. His redeeming grace was his love of dogs. On Good Friday in 683 A.D., he and some friends irreverently took their hounds out to hunt. During the hunt the dogs suddenly stopped their pursuit and respectfully lay down in front of a great white stag. When the stag turned, Hubert saw the image of the cross between its antlers and heard the voice of God telling him that it was time for him to begin to hunt for virtue. Shortly thereafter, Hubert took Holy Orders, established an abbey, and eventually rose to be a bishop of the Church. At the abbey he continued to breed dogs and eventually created the St. Hubert hound, from which our modern bloodhounds are descended.

At this church in North Carolina, the white-robed priest mounted the church steps and gave the "Mass of the Dogs" in honor of St. Hubert. At the end, the oldest dog was called up and blessed, and then every dog in turn received a benedictory touch from the priest. I am certain that no dog present barked even once during the entire proceedings. Actually, according to my recollection, many dogs seemed to bow their heads reverently during the prayer service.

In many divisions of Christianity, patron saints are still commonly called upon to intercede in day-to-day affairs. Patron saints are popular among various interest groups, professions, and even cities, which usually choose a particular saint because of his or her specific talent or because some life event overlaps with the group's special concerns. Sometimes the connections seem a bit tenuous, such as naming St. Clare of Assisi the patron of television because one medieval Christmas when she was too ill to leave her bed she still saw and heard the Christmas Mass— even though it was taking place several miles away. Nonetheless, prayers to a patron saint with whom the petitioner shares some connection are believed to be more likely to be answered. If the saint thinks that the cause and the person are worthy, the saint will intercede with God, asking God to answer the prayer. The reasoning is that since the actual request is now coming from a pure and holy soul that God has already blessed, He may be more predisposed to answer favorably.

Given the somewhat hazy way in which patron saints are designated, I was not surprised to find that there were some differences of opinion among the priests at the wedding as to who the patron saint of dogs was or even whether it was appropriate to have a patron saint for dogs. Young Father Allen began by noting that "We don't do a special mass for dogs in my church. Instead, around October 4, the feast day of Francis of Assisi, we hold a mass for all of the animals, since he loved them all. You might remember that, like Dr. Doolittle, Francis supposedly could understand the speech of animals. Everybody brings their pets then, not only dogs, but cats, hamsters, bunnies, and birds."

Father Rommy was a bit more tentative. "If I had to make a guess, I suppose that Saint Bernard should be the patron saint of dogs. After all, he is responsible for breeding those wonderful rescue dogs at his hospice in the Alps, and also for that famous quote '*Qui me amat, amat et canem meum,*' which translates into 'Who loves me, also loves my dog.' "

Father Francis laughed, and said, "You are mixing up the two Saint Bernards. Saint Bernard of Menthon is responsible for the rescue dogs and hospices, but it was Saint Bernard of Clairvaux who said, 'Love me, love my dog' around a century later. His statement had nothing to do with dogs, however, and he really meant something like, 'If you love me, you must love my faults as well as my good points.'

"I seem to remember that dogs have several patron saints," Father Francis continued. "It is certainly the case that Saint Hubert was one. However, I also remember Saint Vitus mentioned in that regard, but I can't exactly recall the details. It had something to do with the fact that the emperor's dogs wouldn't attack and kill Vitus when commanded to do so, I think.

"A lot of people certainly believe that Saint Roche is the dog's patron. I know a few churches that have a celebration mass for dogs around August 16, which is his feast day. His story goes that he was making a pilgrimage to Rome and arrived in Italy during a period of plague. He tried to comfort some plague victims by praying with them and tracing the sign of the cross on their foreheads. Miraculously these people were healed. He even managed to cure the Cardinal of Lombardy in this way. Anyway, all of this exposure to plague victims eventually resulted in his coming down with the disease, and the ungrateful townspeople drove him out of the city.

"Roche had no one to care for him, so he set up a hut made of leaves and branches, and was blessed by a spring of fresh water that appeared nearby while he slept. He probably would have died in the forest, except that a dog which belonged to a local nobleman found him. The dog licked his wounds, and then ran off. Later the dog came back carrying a loaf of bread. The dog returned each day to clean his sores by licking them, and also to bring him bread. Eventually the nobleman became curious when he noticed the dog stealing bread from his kitchen and dashing into the woods with it. So one day he followed the dog and found Roche. Together the dog and the nobleman cared for him, and eventually his own case of plague was cured by divine intervention. Certainly in statues and paintings Saint Roche is almost always presented as having a dog beside him.

Saint Roche and the dog that kept him alive.

"To the best of my knowledge, however, there is no Saint Fido, Saint Rover, or Saint Sniffer!"

While we chuckled at that, Father Rommy rejoined the conversation. "I am not really clear about the patron saint of dogs, but I did run into a story of a dog that was believed to be a saint when I was visiting France. Saint Guinefort was a greyhound who was a trusted member of a noble household near Lyon.

"The story goes that he was left to guard the noble's young son who was sleeping in his crib. However, when the father returned he found that the crib was overturned and there was blood everywhere, including around the mouth of Guinefort, who came leaping at him in joy at his return. He was horrified and assumed the dog had killed his child, so he took out his bow and killed the dog. As he went over to the crib, he found the child under it and alive, and next to the crib was the body of a snake. The local priest felt that this snake had actually been possessed by Satan, who was trying to use it to steal the child's life and soul. That meant the nobleman had not only killed the dog that had saved his child's life, but a dog who had possibly saved his eternal soul as well. The nobleman was sad and grieved his mistake. He reverently buried the dog and planted a grove of trees around his grave to honor his Guinefort's bravery.

"When people living nearby heard this story, they interpreted the events as if the dog was actually a holy martyr. Because of this they began making pilgrimages to pray at the grove and the grave of 'Saint Guinefort.' According to local customs, it was believed that the dog was a special patron of children who were sick or injured, and a number of miraculous cures were credited to Saint Guinefort.

"All went well until a zealous inquisitor by the name of Étienne de Bourbon showed up with the idea of making sure that no one's behavior strayed from official Church practices. When some women admitted that they had taken their children to be healed by Saint Guinefort, he assumed that this must be some local holy person and felt that he should investigate, since the use of the title 'saint' was probably being applied without Church sanction. As you might have predicted, when he learned that Guinefort was a dog, he was greatly offended.

"The story appears to be real, since I actually got to see the thirteenth-century documents that Father Étienne wrote describing what actions he took next. He had the dog formally designated as a heretic. Acting on that judgment, he had Guinefort's body dug up and the sacred grove of trees cut down and burned along with the remains of the dog. Apparently a dog cannot be an official saint, although he can be an official heretic!"

Chapter 5

Cloning Rover, Fluffy, and Snuppy

THE FOUNDER OF the University of Phoenix, John Sperling, is a multibillionaire, but his feelings for his adopted dog, Missy, were much like those that we average folks have for our pets. In his eyes, Missy was beautiful, talented, and—perhaps unfortunately—unique. As she grew older, Sperling realized that he could never obtain another dog like Missy through the usual methods of breeding, because no one knew what combination of Siberian husky, border collie, and mystery mutt formed Missy's heritage. Furthermore, because she was spayed, Missy herself could not pass on her own special characteristics in the natural way.

Like the rest of us, Sperling wished he could have another ten years with his dog, but unlike the rest of us, he could invest several million dollars to fund university researchers from Texas A&M to find a way to clone Missy.

When word got out about what they called "the Missyplicity Project," phone calls and e-mails poured in from people around the world who also wanted to clone their own pets. In response, Sperling founded a company called Genetic Savings & Clone (GSC) to collect and store genetic material from pets for cloning. Perhaps to Sperling's disappointment, the first cloned pet was a cat named CC (short for Copycat). GSC then began filling cloning orders for cats at $50,000 each (although improvements in technology allowed them to drop the price to half that for each cloned kitten).

Each species presents its own problems for cloning, and dogs are particularly challenging. Cloning starts with a dog egg or ovum, which is released from the ovaries at a very early stage of development and then matures as it travels down the oviducts toward the uterus. Because the eggs are not yet mature, they are particularly delicate, and collecting them is difficult. Once you manage to extract an intact egg, you are looking at a fairly transparent sphere roughly 100 micrometers in diameter. An embryologist places the egg under a microscope, extracts its DNA with a pipette, and replaces it with the DNA from the animal being cloned. The cloned embryo is then implanted in a dog's uterus while she is in estrus (in heat), which only happens once every six to twelve months.

Ultimately, it was not the GSC research team, but Woo Suk Hwang (the South Korean researcher involved in the controversial experiment that supposedly created a cloned human embryo) who was successful. Hwang used this painstaking technique to clone the first dog. The Afghan hound puppy that resulted from this procedure was named "Snuppy," an acronym for Seoul National University puppy. The donor dog provided DNA from a cell from his ear. The surrogate mother that carried the pup was a yellow Labrador retriever, and the birth was by Cesarean

Snuppy (center) with her genetic mother (left) and the surrogate mother (right) who actually gave birth to her.

section. To get some idea of how difficult and chancy the process is, consider that Snuppy was the only one of 1,095 cloned embryos implanted in 123 dogs to survive to healthy puppyhood.

Before you start emptying your bank account to copy your own Lassie or Fido, you should understand a few things about cloning. I am not referring to the ethical and religious objections by some people who suggest that cloning is tantamount to "playing God" or "cheating death." The first thing that you should remember is that a clone of your pet is not a biological photocopy of the original. He or she will not have the same memories, skills, and perhaps not even the same personality as the original. While genetics is important, much of any animal's behavior is shaped by early and continuing experiences. You are not the same person you were ten years ago. You may be living in a different city, with a different family structure, a different work schedule, and so forth. This means that you will not provide the clone with the same environment and experiences that your original dog had, and because of these differences Lassie II won't turn out to be exactly the same as her namesake.

Cloning should really be viewed as a complicated and expensive way of producing an identical twin of your dog—one that is born at a different time and brought forth by a different mother from the original. Cloning is not a way of bringing the animal you love back to life. Not only will the clone not be the same dog you knew as an adult, it is unlikely that your second Lassie will be a perfect copy of the newborn puppy who grew up to be that dog you loved. Cloning does produce animals that are genetically identical to their originals; however, being genetically the same does not ensure that the clone will be physically and psychologically identical to the animal you are trying to copy.

One example of how cloning may not produce the expected results was explored by Ira Glass, host of the radio and television show *This American Life*. The program looked at Ralph and Sandra Fisher, who were the proud owners of Chance, a Brahma bull. This is the same breed of bull used in bull-riding rodeo competitions, and these animals have a reputation for being nasty, aggressive, and unpredictable. However, Chance had such a gentle nature that he was a semi-celebrity by his death at the age of nineteen. It was clear that Ralph loved Chance as much as many dog owners love their pets, and he could not bear the thought of losing such a uniquely mild and mellow example of his breed. Therefore, shortly before Chance's demise, Ralph arranged for Texas A&M University to clone him. They named the clone Second Chance. Unfortunately the clone does not have the original Chance's gentle demeanor. At his fourth birthday party (Ralph has a huge party for his bull every year), Second Chance gored Ralph in the arm. A year and a half later, while Glass and his film crew were visiting, Second Chance attacked again. This time, Ralph wound up with eighty stitches in his scrotum. Still, Ralph remains convinced that once the bull gets testicle-goring out of his system, he will mellow with age and become a loving, tender bull like his predecessor. Unfortunately, despite the fact that Second Chance may be a clone of a harmless and angelic animal, there are no guarantees.

The reasons why clones and their originals can differ is partly biological. The genes that are used to start the clone are taken from existing cells in an adult animal. But adult cells are programmed for directing the activities of skin cells, blood cells, and so forth, not for developing an embryo. So, had researchers tried to clone a dog simply by causing one skin cell to divide without reprogramming its DNA, they would not have ended up with a puppy, but at best a puppy-sized glob of skin cells. To produce a puppy, the DNA must be reprogrammed to exactly the state it was in when the original dog was a fertilized egg. How this reprogramming comes about is a messy and haphazard process that is neither fully understood nor controllable. For this reason, the likelihood that every single gene will be reset to its original state correctly is very low.

The problem is that if some genes are not reprogrammed correctly, the result can be miscarriages and infant deaths, both of which occur in clones. This fact explains why it took so many tries to obtain Snuppy. Clones also often have birth defects and compromised immune systems. For example, in another live birth in the South Korean puppy cloning project, the puppy's immune system was so weak that it died of pneumonia a few days after birth. In fact, shortened life spans and increased likelihood of death by infection are typical problems with cloned animals, starting with the very first mammal cloned, which was Dolly the sheep.

In reality, traditional dog breeders derive the same effects as cloning through selective breeding, which allows them to produce animals that have a particular physical appearance, desired temperament, and genetically determined instinctive behaviors. They have been pretty successful at this. Consider the experience of Kurt Koffka, one of the founding fathers of Gestalt psychology, who always had dachshunds, all of whom were the popular chestnut-colored, smooth-haired variety. Over his lifetime, Koffka had seven such dachshunds—every one of them named

Max. Once he was asked, "Why do you name all of your dogs Max?"

Koffka replied, "The first one was named Max. When he died, I got another, and I gave him a different name. Yet he looked like Max, and he acted like Max, and sometimes I found myself calling him Max. So I said to myself, 'If he wants to be Max, then he is Max.' I suppose that I just wanted him to be Max and to live forever. That is the nice thing about a purebred dog. If you find one you like, you can have him again and again, since they are all so much the same. I suppose that I call them all Max because they *are* all Max, and that is who I wish to live with."

To have these virtual "Max clones," Koffka did not have to spend the estimated $150,000 that it will cost John Sperling to have the Korean lab clone his beloved Missy.

Chapter 6

Venus, Mars, or Pluto?

DO YOU SEEK male or female canine companionship? Does it matter? In human beings, an important issue in choosing our companions (and in deciding how we relate to them) is their sex. Whether we are male or female determines the clothes we wear, the activities we engage in, and even in today's enlightened environment, the occupations and societal roles that are most easily available to us. So it is perhaps not surprising to find that some people have strong opinions about owning male or female dogs, and that in many ways, these opinions reflect cultural attitudes about males and females of our own species.

There is elaborate folklore about differences in the behaviors of men and women, and many people tend to believe that this lore is also true for dogs. Let's consider a few of these beliefs.

Most people think that female dogs make better pets. Dog breeders report that it is quite common for potential buyers to specifically request a "sweet girl." This is based on the widely held belief that male dogs are more aggressive than female dogs. This is really an extension of our beliefs about people—namely, that men are more likely than women to be physically aggressive and dominant, to engage in criminal behavior, and to opt for careers in which they may encounter violence, such as the military or law enforcement.

As with humans, however, science does not find that the issue of sex differences in aggression is simple and predictable when it comes to dogs. Evidence suggesting that male dogs are more aggressive is consistent with the fact that aggressive behavior can be triggered by testosterone, the principal male sex hormone. In dog-versus-dog aggression, male dogs do posture, threaten, and challenge each other more than females, though this is largely ritualized display aimed at establishing social rankings. While such behavior can be disturbing and embarrassing to their owners, serious injuries to the dogs are rare.

Female dogs threaten less frequently, but overall are more independent, stubborn, and territorial than their male counterparts. The females are actually much more intent upon exercising their dominance and while males can forgive an occasional transgression of canine protocol or a failure to recognize their status, females cannot. This explains why fights are more likely to break out between two females, and often without much advance warning. These fights between females are more of a "no-holds-barred" affair than the males' ritualistic fighting, which includes snapping at the air in front of an opponent or using inhibited bites to threaten rather than maim.

In dog aggression toward *humans*, which is of far greater concern, unneutered males are more likely to be involved in biting incidents. Furthermore, since male dogs are larger, such bites can be more serious. Most human-bite incidents by pet male dogs involve leadership and control issues between the dog and the human. But, just as in dog-versus-dog aggression, there is usually ample warning and many threats before anything physical occurs. Both male and female dogs are equally likely to threaten or bite if the issue is possession of a food or a cherished object. A female is less likely to wage a dominance battle that escalates into aggression with a human, but females can be cunning and resourceful in getting their own way, and in the human mind they are often perceived as more "sneaky" than male dogs.

Female dogs that have just given birth will aggressively defend their puppies from anything that might threaten them. This is a completely unrestrained use of force, since a canine mother will do anything to protect her litter. Early socialization to a variety of different people will reduce the likelihood of such aggression when the female has a litter.

Unfortunately there is an additional, little-known complication in female dogs that seems to set them apart from wild canines and other domestic animals. Whether pregnant or not, after ovulating all unaltered female dogs go through a two-month period in which their body is flooded with the same hormones that are present during pregnancy. For some dogs this will result in physiological changes that mimic pregnancy, such as lactating. The problem is that in the last three or four weeks of this phantom pregnancy, the female may start acting in a strange manner about certain items, such as tennis balls, socks, soft toys, or shoes. Typically she collects these and hides them under a bed or other piece of furniture. Furthermore, she might become possessive and protective of these items and snap, growl, or bite anyone who comes too near or disturbs them. Like dogs with actual litters, however, well-socialized dogs are less prone to this behavior. But just as in the case of male aggression, the only real preventive measure is early neutering.

If this problem does occur in your female dog, behavioral

methods won't cure the aggression. Hormone treatments can eliminate it, or you might simply wait the situation out, since it will usually disappear in a few weeks by itself. However, during the time that this form of aggression is likely, isolating the dog might be best. At the very least, keep children or non-family members from approaching that pile of toys, since the dog is treating these as her litter and will protect them as fiercely as she would her real puppies.

Because a lot of the behaviors we have talked about are controlled by hormones, neutering can take sex out of the picture by removing the sex hormone-producing apparatus—the testicles of males and the ovaries and uterus of females. The male hormone testosterone influences territorial urine-marking, dog-to-dog aggression, and the tendency of male dogs to roam more than females (in order to stake out territory and look for a mate). Female sex hormones, on the other hand, only affect a dog's personality during her heat cycles, which usually occur twice a year, when she is most likely to urine-mark and wander. Progesterone, a hormone involved with the female cycle, has a generally calming effect, but also stimulates a possessive or protective

For several weeks after coming into season, a female may guard her toys and possessions as if they were her litter of puppies.

attitude toward her puppies or puppy substitutes, such as toys or young children. Neutering stops this twice-yearly potential for aggression.

Neutering is not a cure-all. It rarely affects fear-biting, territorial aggression (the dog's natural defensive reaction when something comes near his home), or predatory aggression (the tendency to chase things that run, and to nip or bite them).

Neutering does affect the dog's personality quite subtly in other ways. Neutered dogs seem to pay attention more to people because they are paying less attention to sex-related activities of other dogs. In addition, neutering a dog that is not yet an adult seems to freeze personality development at that stage, at least in terms of keeping certain puppylike traits in place. Since the optimal time to neuter a dog is just before puberty, this becomes a useful tool if you have a dog breed where the adult displays pronounced aggressive tendencies. Generally speaking, the puppy is softer and less likely to show dominance and other tendencies that can lead to aggression later. So, if you have such a breed and you like the pup's personality at the age of six months, then, subject to your veterinarian's approval, you have a good reason for neutering at that age. For the average family dog, socialization and training are the most important factors for preventing or curbing the development of aggressive behavior. However, neutering certainly helps.

Another common belief is that male dogs bond better with women, and female dogs bond better with men. This idea seems to be an extension of the simple heterosexual belief that men are best suited to live with women and vice versa. The truth of the matter is that male dogs are more likely to bond strongly with one person, while females form much looser relationships—although breed differences again are important in this matter. The sex of the person seems to have little effect on these behaviors.

A probably even more commonly heard statement is that female dogs are more loving and emotionally responsive than males. This is a simple extension of our societal stereotype that women are more "motherly," warm, and supportive. Women are also imagined to be more moody, more easily stirred to tears, and subject to anger. To get the real answer, we have to separate the issue of "warm and affectionate" from moodiness. Female dogs are much more prone to mood swings, just as the stereotype gleaned from human folk psychology predicts: one day the girl may be sweet and affectionate, seeking your company, while on another day she may be reserved, withdrawn, grumpy, and unsociable. Males tend to be more steady, steadfast, and reliable. Their behavior tends to be more predictable and less moody. Having said this, we have to remember that the breed of a dog is much more important than its sex in determining emotional stability. Basset hounds, bulldogs, and Newfoundlands are generally placid in most situations, whereas Chihuahuas, dalmatians, and most terriers tend to be more excitable and reactive. Sex does make a difference, but not at the level predicted by folklore. In emotionally excitable breeds, females tend to be more volatile and reactive than males. In emotionally stable breeds, the sex differences are much smaller.

The truth about sociability and affection differs from the folklore. Generally, females can be sociable and will often come to you hoping for some attention. However, when she's had enough, she will move away and attend to other business. In contrast, males are usually consistently more sociable, affectionate, exuberant, and more demanding of your attention. Males are very attached to their people, but are also more outgoing. Generally, males are more likely to accept other pets and to bond more quickly with children. Females do seem to be more attentive to very young infants, perhaps because their pheromones—special scents produced in animals—are very similar to those of puppies.

Male and female dogs tend to age differently, which affects their personalities. Most males tend to keep more of their silly, soft, puppylike behaviors longer than females. Thus males tend to want to play games more, and remain fun-loving till the day they die. Females, however, tend to become quieter and more reserved or dignified as they age. As I think of it, perhaps there is a corresponding human stereotype. Television programs love the character of the old grandfather who still talks, plays, and noisily interacts with the children, while the quiet grandmother works in the kitchen. She only joins the activities when she thinks things are getting out of hand and everyone needs to quiet down a bit.

The truth is that both male and female dogs are good pets. Overall, preference for one sex or the other in dogs is mostly a matter of human inclination and personal notions. For example, one woman informed me that she had always had female dogs because "I don't want some boy dog leering at me as I dress and undress." On the other hand, I heard another woman explain that she always kept male dogs because, "If someone tries to attack me or break into my house I want a man around, not a girl dog who will be as frightened as I am."

In one case, the sex of a dog had great historical significance. Frederick the Great, Emperor of Prussia, always had female Italian greyhounds as his companions. When he built his mansion, *Sans Souci,* he said that no females, other than his dogs, would ever be allowed into his bedroom or the private wing of the residence again—not even his wife. At one dinner party he hosted (with no women present), talk turned to the state of the French court under King Louis XV. The king's relationship with his mistress, Madame de Pompadour, had become an item of gossip. Born Jeanne-Antoinette Poisson, Madame de Pompadour was a brilliant woman, politically astute, well cultured and ambitious, who had attracted the attention of the French king. After she became Louis's mistress, she was awarded the title of Marquise de Pompadour, after the name of the elegant manor the king gave her. Madame de Pompadour got to know everyone of influ-

Frederick the Great and Biche.

ence at the court and was always aware of important political activities. Accordingly, the king came to rely heavily upon her opinions, and she functioned more like a powerful executive member of the government than merely his mistress.

At this particular dinner party, Frederick, with typical cynical and cutting wit, gestured to a favorite dog, Biche, who was sitting beside him. "This is my Madame de Pompadour," he said. "She sleeps in my bed and breathes her advice in my ear. The only difference between my Pompadour and that of Louis is that he has conferred upon his the title of 'Marquise,' while I have conferred upon mine the title 'Biche.' " The group laughed heartily. "Biche" is the French word for "bitch," which even then had the dual meanings of both a female dog and a lewd and wanton woman.

Unfortunately, word of this incident made its way back to Madame de Pompadour. She was enraged by what she took to be humiliating and demeaning treatment. In revenge, she set her mind to turning Louis XV against Frederick and Prussia. Largely because of her influence, instead of forming an alliance with Prussia, France sided with Frederick's traditional enemies, Austria and Russia, the result of which was the Seven Years' War.

So, while the sex of your dog (especially if neutered) might not make a great deal of difference to your life, what you choose to say about your dog could have great import. As always, it is the affairs and attitudes of humans that seem to predominate.

Chapter 7

Do You Look Like Your Dog?

AT A GATHERING of dog owners, I found myself sitting at a table with a woman who was explaining, "You can always tell which member of a family actually picked the dog. People always pick dogs that look like themselves. If you want a well-known example of this, just take the case of Winston Churchill. Now there was a man who certainly looked like his pet bulldog!"

Winston Churchill was of course the British prime minister who helped guide England through the dark days of World War II. The fact that Churchill in his mature years looked like a bulldog is indisputable. The round full face, the flattened features, the jowly cheeks, the large wide mouth, the skin folds around the eyes: all looked much like those of the English bulldog. In addition it is interesting to note that if the woman speaking about Churchill had had long blond hair, that paired with her tan-colored pants and vest certainly would have shown a strong resemblance to her own dog—a golden retriever.

Actually, there are scientific explanations for why people might prefer dogs that look something like themselves. The data are clear that, at least when dealing with people, we certainly prefer those who are similar to us. The success of computerized dating services, for instance, relies simply on matching on the basis of similarity. These services give extensive questionnaires in order to find out basic information about their clients, including their religion; their family's social status and income; their political beliefs; their taste in music, entertainment, and sports; and so forth. Then they match people on as many dimensions as possible. The better the match, the more likely that people will develop a mutual attraction.

The interesting thing is that research shows that if you want to predict whether people will like each other, you tend to do better if you include some aspects of the person's physical appearance, along with those measures of attitudes and personal history. How tall a person is, how much he or she weighs, hair color and skin tone will all have an influence since people prefer others who share their physical characteristics. While most people like physically attractive people, the most successful matches pair people of about the same degree of attractiveness. Thus beautiful people are happiest with other beautiful people, average-looking people with average-looking partners. So even had he survived, it is highly unlikely that in real life, the ugly, misshapen Quasimodo of *The Hunchback of Notre Dame* would have ever ended up living happily ever after with the beautiful Gypsy girl Esmeralda.

There is some evidence that people tend to select dogs based on how similar that breed's personality is to their own. For example, Humphrey Bogart was known for the strong, rugged roles he played in motion pictures like *The Maltese Falcon, Cas-*

ablanca, and *Treasure of the Sierra Madre.* Director Howard Hawks once commented that "Bogey thinks that he has to live up to the reputation of all of those tough guys he plays." Hard-drinking and rowdy in his personal life, he always owned dogs with tough, self-sufficient characters, such as boxers and Scottish terriers.

Contrast this to another classic film actor, James Stewart. The American Film Institute observed that Stewart was an actor "so beloved by the movie-going public that they call him 'Jimmy,' just like a member of the family." He had a real-life personality similar to the warm, down-to-earth, friendly men that he played in films like *You Can't Take It with You, Mr. Smith Goes to Washington,* and *The Philadelphia Story.* Stewart surrounded himself with dogs of the same temperament, namely golden retrievers.

You might feel that it would take a leap to get from how we might select a dog that has a personality like our own to the conclusion that we might choose a dog that looks like us as well. Yet there is a way to get there using a psychological mechanism that is subtle yet simple, namely *familiarity.* Simply put, we like things that are familiar. This explains why we are so willing to read or view each new version of the King Arthur legend, or why people go back, year after year, to hear the same opera, or why radio stations that play only "oldies" are so popular. Advertisers repeat the same ad so many times because they know that "repetition builds reputation." It also explains why people vote for actors and the sons, daughters, or wives of well-known people without any knowledge of their actual competence for the elected position—it is simply because the name is so familiar that a positive feeling has grown up around it.

One scientist demonstrated this in an amusing way. He showed unilingual English-speaking people a series of Chinese characters, without their translations. When the people were later asked to guess what these characters actually meant, the ones that had been shown a number of times (so they were now familiar) were more likely to be "translated" by people as meaning something positive and favorable.

Scientifically, we have now reached the important part of the story—your face. We all are quite familiar with our own face. We see it in the mirror every morning as we shave, put on makeup, or comb our hair. We see images of our faces thousands of times each year as we pass by various reflecting surfaces in the environment. Science therefore suggests that, as in the case of everything else that we have seen many times, we should be rather fond of the look of our face. It is also likely that we will also transfer some of that sentiment to anything similar enough to remind us of our face.

Some psychologists have argued that this explains why children who look very much like one of their parents tend to be favored and treated more lovingly by that parent. It might also provide a link to why people end up with dogs that look something like themselves. If the general features of a breed of dog's face resemble the general features of our own face, then, other things being equal, that breed should arouse a more warm and loving response on our part.

Since there had not been much scientific work done on the resemblance of dogs and their owners, I tested 104 women students enrolled at the University of British Columbia. First, they were shown slides containing dog portraits of four different dog breeds. Each portrait was simply the head of a dog looking toward the camera. The four dog breeds included an English springer spaniel, a beagle, a Siberian husky, and a basenji. For each dog, the women simply rated how much they liked the look of the dog, how friendly they thought it was, how loyal they thought it might be, and how intelligent it appeared to be.

Afterward, I asked the women some questions about their lifestyles and also asked them to look at a series of schematic sketches of hairstyles and to indicate which was their own most typical style. I was not interested in details of their coiffure, but only in certain general characteristics. Specifically I divided these hairstyles into two groups: the first contained longer hairstyles that covered the ears, while the second group contained shorter hair or longer hair that was pulled back so that the woman's ears were visible.

In general, women with longer hair covering their ears tended to prefer the springer spaniel and the beagle, rating these breeds higher on the dimensions of likability, friendliness, loyalty, and intelligence. Women with shorter hair and visible ears tended to rate the Siberian husky and the basenji more highly on these same dimensions.

The reason for this result may have to do with familiarity's effect on preference. Longer hair on a woman forms a framing effect around her face, which is much the same as the framing effect created by the longer, floppy ears of the spaniel or beagle. Shorter hair gives more visible, unframed lines to the sides of the woman's face and allows her to see her own ears. Both the Siberian husky and the basenji lack the drooping ears that frame the face like long hair, and both have clearly visible pricked ears. Obviously, we are not talking about an overpowering effect on preference, since there were a number of women with short hair who preferred the long-eared dogs, and some long-haired women who preferred prick-eared dogs. However, the size of this effect is large enough to be statistically reliable and could confirm the common belief that we look like our dogs to some degree.

Given that hairstyle variations of this sort are only sensible to talk about in women, this particular research model was somewhat limited, so Michael Roy and Nicholas Christenfeld, psychologists from the University of California at San Diego, decided to extend my research using another technique. They photographed forty-five dogs (twenty-five purebreds and twenty-

Long hair frames the face much like lop ears, while visible ears give an effect much like prick ears.

two mongrels) and their owners, separately. They then showed photos of the owners to twenty-eight volunteers and asked them to guess which was most likely that owner's dog from a pair of pictures. Each pair contained the dog actually owned and another dog. A dog was regarded as resembling its owner if a majority of the volunteers matched the pair. The volunteers were able to match purebred dogs with their owners correctly in about two thirds of the cases, which seems to confirm that dogs and owners look alike.

There was an interesting quirk in the data, however. There was no link between the appearance of mixed breeds and their owners. Coresearcher Christenfeld thought that this was reasonable.

"When you pick a purebred, you pick it specifically because of how it's going to look as a grown-up," he pointed out. "On the other hand, mutt owners, like me, make our choice on the

Winston Churchill with an English bulldog (left) and his daughter's pug (right).

spur of the moment at a dog shelter. The truth is that we really don't know what the grown puppy will look like."

Let's go back to the case of Winston Churchill, whom many people claim as a perfect example since he clearly resembled his pet bulldog. The problem is that Churchill did not own a bulldog. The British often consider the bulldog as the symbol of their country, and Churchill was the leading political figure in Britain. These facts are what probably led to the erroneous conclusion in the public mind that the great man also owned a pet bulldog! Churchill's own dog certainly would not have been selected on the basis of visual similarities. If you could have looked into Churchill's bedroom, curled up around his feet on the bed you would have found a poodle named Rufus. With his narrow, pointed muzzle, clean unwrinkled face, and close-set eyes, this dog did not look even faintly like his master. This breed of dog was not merely some accidental match since, when the original

Rufus died, the great man replaced him with another that looked virtually identical to the first one and had the same name. "He is named Rufus II—but the 'II' is silent," Churchill explained.

Still, despite the case of the prime minister and his poodle, research does seem to show that dogs and their owners resemble each other more likely than not. And in Churchill's case—well, his daughter owned a pug, which might be close enough!

What Dogs Do

Chapter 8

Dogs in the Witness Box

IN PHILADELPHIA IN 2002, a postman opened a mailbox and discovered a package that appeared to be some kind of explosive device. The Philadelphia Police Bomb Squad was called in, verified that it was a bomb, and defused it. The next day, in a mailbox about a mile and a half away, a similar package was found, and again the bomb squad was needed to render it safe.

Whenever a crime is committed, evidence is always left behind. Of course some criminals may be cautious enough to avoid leaving behind obvious clues, such as fingerprints, hair, or other forms of trace evidence. However, despite the best efforts of the most meticulously careful criminal, one type of invisible trace evidence is almost always deposited at the scene of a crime and on the items they used as part of their nefarious activities. No matter how hard they try, offenders cannot avoid leaving behind their unique human scent.

Two days later, with no other source of evidence as to the maker of these bombs, the police decided to use a Scent Transfer Unit to gather any remaining scent evidence. This is a portable device that first creates a vacuum around the evidence and then uses airflow to carry any existing components of human scent into sterile gauze pads. These pads can be stored in ziplock or heat-sealable bags until needed. This is much better than simply wiping the objects with gauze pads since that kind of physical contact may disturb other potential sources of evidence, such as fingerprints or traces of DNA.

A bloodhound was taken to one of the mailboxes along with a scent pad prepared from the bomb. The dog was allowed to sniff the pad and then told to track. Although any dog can track by scent, some dogs are considerably better at this than others, and the ultimate canine nose is that of the bloodhound. A dog's nose is many hundreds of times more sensitive than a human nose, which has on average about 5 million scent-discriminating cells. A German shepherd has 225 million of these olfactory receptor cells, but the bloodhound's massive nose has more than 300 million, making it the most sensitive nose in all of dogdom. Some scientists suggest that bloodhounds may be sixty times more sensitive to scents than the German shepherd, which is already a thousand times more sensitive to smells than we nasally-challenged humans. According to legend, the bloodhound was first bred around 1000 A.D. from tracking dogs used by the monks at the St. Hubert Monastery in Belgium, with contributions from some French hunting hounds and the English Talbot hound. Records that trace back to the sixteenth century report that bloodhounds were already being used at that time by law enforcement officials to track thieves and poachers.

For scientific procedures to be accepted by courts as evidence, they must be tested and validated, and something about their accuracy and potential susceptibility to errors must be known. With this in mind, the Federal Bureau of Investigation began to sponsor research into the accuracy of bloodhounds doing police work. In 2001 the FBI and the Southern California Bloodhound Handlers Coalition jointly conducted a study to determine the usefulness of human-scent evidence on bomb components and arson devices. The investigators decided to place some pretty heavy demands on the dogs because they were interested in more than just the accuracy of scent evidence from those lucky incidents where the explosive or arson device was detected before it was detonated; they also wanted to know if such traces of scent would survive explosion and fire in instances where the device had already exploded. If the dogs could do this then they could be a valuable asset in investigating terrorist bombings and deliberate acts of arson.

In this particular study, twenty bloodhound teams of varying ages and experience were used. Some had been previously trained on bomb debris, others on arson debris, and the remaining dogs had been trained in the traditional manner of simply following any designated scent. Four pipe bombs, each with a different type of explosive, and two gasoline-based arson devices (one metal and one plastic) were prepared. Each was then handled by different individuals who were the stand-ins for actual criminals. Afterward, the bombs were detonated and the arson devices set off and allowed to burn for two minutes before being extinguished with water. Once the smoke had cleared and things had cooled off enough to be safely handled, scent was collected on sterile gauze pads using the scent-transfer device.

Two weeks later, test stations were set up at El Dorado Park in Long Beach, California, a park used by the public, meaning that the distracting scents of joggers and other park visitors would be present. The six individuals who handled the explosive and arson devices set up several trails, each walking away from the test station alongside another person who had not touched any of the devices. At some distance along the way, the two people split off in different directions to force the dog to make a choice. The bloodhounds in the test were brought in after the trail had been made. Each had to sniff one of the collected scents from the explosive devices and then was given the command to track. In 78.3 percent of the cases the dogs indicated they recognized that the scent from the explosive device matched the scent of someone in the immediate area. In 88.6 percent of those cases where they knew the scent was present, the dogs were able to successfully follow the trail to the place where the person who had handled the device was hiding. One piece of data that indicates just how accurate the dogs are when doing this kind of work is that there were no false positives (that is erroneously identifying a person whose scent was not collected).

While most dogs are unable to track a scent trail that is older than twelve hours, researchers have shown that experienced bloodhounds are able to identify scent trails after forty-eight hours with a success rate of up to 96 percent with no false positives, even when the tracks have been contaminated by being crossed or mixed with other people's scents. If the scent buildup is strong enough, dogs can make identifications over periods of months. The FBI demonstrated this in 2003 when a subject's primary residence was located by scent. The test subject was a woman who had lived in a house in Stafford, Virginia, for seven years then moved to New Mexico. Six months after she left, she mailed a letter to the FBI, which traveled through the U.S. postal system from Albuquerque to Stafford. The scent was then lifted from the letter, and a bloodhound team was started at an intersection several houses away from the woman's previous residence. The bloodhound indicated a matching human scent, and then trailed it until he identified the house in question. This was all done despite the fact that the sample scent must have been heavily contaminated with other human scents after passing through the mail, and the house had not been occupied by the woman who was the scent target for six months.

Given this kind of research, it is not surprising that in the case of the Philadelphia bomb maker, when the dog was given the scent collected from the explosive device and started near one of the mailboxes, he immediately indicated that he had found traces of that same scent nearby, and he began to track it. Even though there had been two days of car and pedestrian traffic in the vicinity, the bloodhound was able to follow the track into a neighborhood where the trail ended. A second dog was given the scent and started a swing through the neighborhood. The dog ultimately identified a house occupied by Preston Lit. After the preliminary identification from the dog, the criminal investigation eventually led police to the same location. When he was arrested, Lit pleaded guilty and was sentenced to serve a federal prison term.

Based on cases like that of the Philadelphia bomber, some police forces have begun to use location-checking by dogs to aid their investigators. When an investigation produces a list of potential suspects, scent samples are taken from them. A dog then determines whether particular suspects have been in the specific residences, businesses, or other areas associated with the crime. This technique has the potential to focus an investigation and rule out suspects; however, such evidence has to be used in conjunction with other investigative techniques. While a positive match by the dog is fairly significant for investigators, the lack of a scent match is not a guarantee that a suspect has no association to a crime any more than the absence of fingerprint evidence guarantees that a suspect is not guilty. When a dog does not indicate a scent match, this may only indicate that this particular odor lacks a strong enough association with the location for the dog to give an alert.

An even more powerful procedure is now being used with bloodhounds to identify criminals. This is called a *scent lineup* since it works much like the usual police lineup, where witnesses are called in to see whether they can pick out an individual that they recognize as being associated with a criminal event from among a set of other people known to be innocent. With a scent lineup, however, a dog tries to match the scent of an object associated with the crime to the scent of a particular suspect.

Unfortunately, dogs cannot be sworn in as witnesses and verbally identify defendants in court. Some courts also are hesitant to accept evidence from canine identifications. For this reason formal procedures for scent lineups have been worked out by the Dutch and Polish national police forces, and evidence from this kind of test has been accepted in many courts in North America. The lineup starts with the object from the scene of the crime that is believed to carry the criminal's scent, or an absorbent pad from a scent-transfer machine that has been used on such an object. Stainless steel tubes are used to carry the scent of the people who will appear in the lineup. The suspect and six "foils" (other people who are not associated with the suspect) are each given two tubes and required to hold them for about five minutes. One of the six foils is designated as the "control." This control person is given an object or a scent pad similar to that which carries the criminal's scent, and it is slipped into their pocket.

To meet the stringent requirements of evidence by criminal courts, the actual testing is slow and methodical and usually videotaped to make sure that evidence based on the dog's judgment is unbiased and certain. First, an assistant prepares the room for a scent lineup by making two rows of scent tubes, each with the suspect's steel tube, plus one from the control, and one each from the five other foils. These are clamped to a platform in random order. The first part of the lineup is really a test to make sure that the dog is working well and can do the job. When the dog is brought into the room, he is first allowed to sniff the object or pad with the control person's scent. The dog is sent to search the first row for a matching scent. If the dog identifies the scent tube that was handled by the control person, then the procedure is repeated with the second row. If the dog identifies the odor of the control person this second time, and if the dog has not shown particular interest in the scent tube held by the suspect, then it is assumed that the dog knows what he is doing and the test will be fair.

The suspect identification stage now begins. At this point, since the dog has retrieved both tubes held by the control person, six tubes remain on each platform, one handled by the suspect and five by the foils. The dog is now brought back to the first row of tubes and is allowed to smell the real criminal scent evidence. If the dog now picks out the tube carrying the odor of the suspect in both rows, then police conclude that the scent evidence object and the suspect share an "odor similarity." Typically such scent lineup evidence is treated by the police and the courts as carrying the same weight as the usual lineup identifica-

A bloodhound inspects odor-carrying cylinders in a scent lineup.

tions and, combined with other evidence, it may lead to the identification of the guilty person. So, in effect the dog gets to sit in the witness box with much the same status as an expert witness offering an opinion of evidence.

Scent lineups have proven to be extremely useful. In one drive-by shooting, the police had little to go on besides the shell casings from the shooter. These were used to produce the scent sample. In a scent lineup this sample was matched to the scent of a suspect. Confronted with this evidence, he confessed and was later sentenced to a jail term.

With scent lineups as their model, scientists have adapted this method to solve conservation problems. Researchers are monitoring the critically endangered Siberian tiger (now often referred to as the Amur tiger) using scent samples from the animals' scat. These secretive animals are difficult to find because they conceal themselves in dense Russian forests. Furthermore, as hunters they often roam over a wide range. Nonetheless, finding tiger droppings was relatively easy since these large carnivores leave scat on trails and in other open places to mark the boundaries of their territories. Tiger scat also has a distinctive shape and never contains any plant material. The hardest part of the scientists' work involved putting together a reference collection so that they could identify individuals. To do this researchers used a combination of tracking and scat-sniffing. They started by first tracking tigers in the snow where they left footprints. Footprints are useful because they allow researchers to determine a tiger's sex by measuring track sizes (males have larger paws). The presumption that the scientists made was that any scat near the tracks was made by the animal that left the footprints. Next, the investigators used dogs to determine whether scat along tracks in one location matched any of those along tracks in other locations.

To identify individual tigers, researchers use the same procedures that police use in a scent lineup, except that the control sample is not used and the scent-target samples are arranged in a circle. The dog first sniffs the unknown scat sample and then sniffs its way around the ring to see if it matches one of the reference samples from known tigers. By mapping the places where scats from particular tigers were collected, researchers can determine how many tigers are living in a given area, where they are going, and whether the population's size is changing. Keeping account of the tigers using scat-identifying dogs is cheaper, faster, and safer than other methods such as capturing live animals using anesthetic darts and implanting radio-monitoring devices.

Researchers have found that adapting other aspects of scent identification that work for criminal investigations is less successful than the scent lineup. For example, taking the dogs into the forest to give them the scent of a particular tiger would be similar to the location checking and tracking that is used when investigating human crimes. Unfortunately, this is not practical because the tigers will eat the dogs. "We've found that the tigers will go out of their way to catch a dog," says biologist Linda Kerley, who coordinates the tiger monitoring project. Thankfully, human criminals have not hit upon the same "catch-and-eat" strategy to avoid having their scent identified by police dogs.

Chapter 9

Why Do Dogs Have Wet Noses?

"ICK!" MATEO, ONE of my young grandchildren, squeaked in surprise. He had been standing next to me watching my computer screen while I adjusted the colors on an image that I needed for a lecture I would be giving the next week.

"Dancer's nose is all wet and cold!" he said. My orange-and-white Nova Scotia duck tolling retriever had come over to my desk to greet my little guest, or at least to check to see whether Mateo was carrying something edible that might accidentally drop on the floor where an opportunistic dog has a chance to gobble it down. The touch of Dancer's nose on his hand had caught my grandson's attention and Matty was now standing with both hands behind his back to prevent any further contact.

Matty then looked up at me and asked the question about dogs that I have heard from children more frequently than any other, "Why do all dogs have wet noses?"

The scientist and professor in me always wants to immediately launch into an explanation based on facts and data. However, at five years of age, Matty was entitled to something more than a lecture on dog anatomy and behavior, so instead I told him the story that I was told at his age.

Back in biblical times, God told Noah that a great flood was coming and that he should build an ark. He was to put his family on the ark to keep them safe from the water. In addition he was to gather up every kind of animal he could find and bring pairs of them onto the ark. These people and animals would become the parents of the next generation of living things when the flood was over, because only passengers on the ark would survive the coming flood.

While Noah and his older sons worked to construct the ark, his younger sons and daughters started to gather up the animals. The first animals brought to Noah were a pair of dogs. He looked at them and said, "You are reliable and intelligent animals. I need some help in watching over things. I want you to make sure that none of my neighbors steals any of the precious wood we need for building this vessel or any of our supplies. And I need you to warn me if something goes wrong. Will you do that for me?"

Since dogs had been created to love humans and would try to do anything that people wanted, these dogs happily agreed and began to patrol the area and keep watch.

Other animals soon came, and those that could help did the best they could. Elephants and horses carried supplies to the ark. Giraffes used their long necks to lift things up to Noah and his sons when they were high on their ladders. Cows gave Noah milk, and chickens gave him eggs, while monkeys climbed high into the trees to gather fruits and nuts to eat once the ark was launched. All the while, the dogs watched vigilantly, warning Noah whenever there was a problem.

Soon the rain began to fall and Noah loaded everybody onto the ark. Noah depended quite a bit on the dogs during this time. Many of the animals were not very bright and kept forgetting why they had gathered together. For instance, the sheep, cattle, and even the ducks and geese often wandered off into nearby fields. With the rain falling it was the dogs' job to find these strays, herd them back to the ark, then up the ramp and into their pens. There was a lot of bleating and mooing and clucking in protest, but the dogs were good at their jobs and every animal got on board before the water rose high enough to be dangerous.

The dogs continued their job of patrolling and watching for trouble while the ark floated aimlessly on the water that now covered the earth. As time went on the rain stopped and the level of the water seemed to be dropping. So Noah sent out a dove to see if she could find dry land. Since the ark was not a pleasure boat but was built to protect its passengers from a hard-driving rain, there was not much of a deck to stand on. Therefore Noah sent the dogs out through a hatch and had them stand on the roof of the ark and watch for the dove's return. Because of that, the dogs were the first to see the white bird returning with an olive branch in her mouth, which meant the waters were falling and the trees were now above the flood tide. Thus it was the joyful barking of dogs that told everyone that their long journey was coming to an end and they would soon be able to walk on dry land again.

However, that happy ending to the story of Noah's Ark might never have happened had it not been for the heroism of the dogs. One night when the dogs were patrolling the ark, they heard the sheep, cows, and horses begin to make frightened, complaining sounds from their quarters in the lowest deck of the ark. When the dogs reached the bottom level, they saw that the floor area was filling with water. The dogs quickly searched the area and found the source of the trouble: a neat round hole, about the size of a twenty-five cent coin that had opened up in the side of the ark allowing water to flow through it.

The dogs looked for something to stop the flow and grabbed some of the hay that the animals had been eating, chewing it into a wad and trying to plug the hole with it. They pushed it as tight as they could with their paws, but because the hole was so far below the water line, the pressure of the water was just too great and their makeshift stopper popped out.

One of the dogs motioned to another to go get help, and as she dashed for the stairs the dog that remained recognized that it would be a while before help actually arrived since the humans all lived on the top deck. The stream of water had to be stopped. He braced himself against the side of the sheep pen to keep from being pushed back by the force of the water, took a deep breath, and shoved his nose into the hole.

The water pressure was very strong, and the water kept pushing the dog's nose out of the way. Although he was able to hold his place, some of the water worked its way into his nostrils and dribbled down into his mouth, making it hard to breathe. But with his hind legs pushing at the slats of the pen he managed to slow the stream of water to a trickle.

No one knows how long it took Noah and his sons to recognize the alarm and follow the other dog down to the bottom deck. Certainly it would have been many long minutes for the people to work their way along the whole length and depth of the ark to reach the leak. By the time they got there, the dog with his nose in the hole was in great pain and barely conscious. When Noah arrived he moved the dog away as carefully as possible; the dog himself was so weak that he nearly fell over. Then Noah and his sons hammered a great plug into the side of the ark to stop the water, and carefully sealed the side of the ark with pitch to make sure that there would be no new leaks.

Afterward, even with everyone working together and bailing water, it would be several days before the lower deck was dry again. It was, however, quite clear that had the dogs not acted so swiftly and heroically, there would have been so much water in the ark that it might have foundered and sunk.

When God saw what had happened, He decided that no one should ever be allowed to forget how loyal, faithful, loving, and clever dogs were. So he awarded dogs a special badge of courage in the form of a wet nose. He chose that particular symbol so that everyone would remember that cold water had nearly wiped out everything alive on the planet, and would have had it not been for the courage and tenacity of dogs. So, since that time, all good and brave dogs have had wet noses.

As I finished my story, my granddaughter Cora, who had apparently wandered into my study, said with authority, "Well, I heard that a dog with a cold wet nose is healthy and one with a hot dry nose is sick." Cora was at that time ten years of age, going on twenty-eight.

Since a ten-year-old can understand facts and science, I took the opportunity to be a bit professorial.

"Actually, Cora, there are two reasons why a dog's nose is usually wet. One has to do with sweating. When you get hot, sweating cools you. Put the tip of your finger in your mouth and let it get wet. Now wave it back and forth so that the water evaporates. See how cool the tip of your finger feels? Well, that's why sweating cools you.

"People sweat all over their body, but dogs can only sweat—and let the evaporation help to cool them—from the pads of their feet and their nose. They can also open their mouths and pant, so that the evaporation from their tongues will help cool them as well. So a wet nose is just part of the way dogs keep cool, since they live all of their lives in fur coats.

"Another important reason is that dogs have a very sensitive sense of smell, but in order to smell things they have to collect

A dog that is trying to smell something may lick his nose to make it wetter.

the chemicals that float in the air on the moisture of their noses. Just as a wet cloth gathers more dust than a dry one, a dog's wet nose catches molecules from the air and makes it easier to smell them. When a dog is trying hard to smell something, he'll lick his own nose to make it wetter so that he can catch more molecules near his nostrils and smell things more easily."

I stopped at that point because I did not want to go beyond the comprehension of a ten-year-old. However, the scientist in me wanted to go on. I really wanted to point out to her that a wet nose is good, but too much wetness is counterproductive for dogs. It works like this: scent molecules are eventually inhaled into the nose. The scent-detecting cells in the nose have little hairlike structures designed to register the chemical makeup of these molecules. That can only be done if those molecules dissolve, so these scent-detecting cells are covered in a thin layer of mucus, which is wet enough to dissolve the molecules but viscous enough to hold everything close to the hairlike detectors.

The next step is to flush the dissolved molecules out of the nose so that the next new scent can be detected. For this reason an average-sized dog may produce more than a pint of thin slimy mucus each day. In most dogs you don't see the mucus that is continually washing out of the nose since it flows down the dog's throat. In some hounds and mastiff-type dogs, however, much of the mucus leaving the nose slides down their inner cheeks, and may spill out of their mouths in the form of drool. Unfortunately, a dog that sheds droplets of drool every time it shakes its head or simply stands quietly in one place is not desirable in today's modern world, where dogs are brought indoors and wet spots on the carpets and furniture are not appreciated. So many dogs with this characteristic have suffered in popularity in recent years, despite the fact that this group of dogs includes such mellow and sweet breeds as the bloodhound and the Newfoundland dogs.

This controlled wetness provided by mucus in the nose works quite well in keeping the smelling apparatus finely tuned in most dog breeds. However there is an exception. Dogs with very flat faces, like pugs or Pekingese, have a problem. This is because their noses are very short, despite the fact that they are still producing the same amount of mucus as other dogs their size. The result is that there is just too much fluid to clear out of the nose quickly, so some of it backs up and actually blocks the entry of some of the scent molecules or does not move away fast enough to allow new scents to be caught and dissolved in fresh mucus. That is why tests have shown that flat-faced dogs are somewhat poorer at discriminating smells than their larger-nosed colleagues.

I was trying see if there was some easy way to convey this information to Cora and Mateo when their father, Geof, my son by marriage, poked his head into the room and asked, "What's up, guys?"

Matty announced proudly, "Dancer has a wet nose because he's a brave dog," once again proving that the glue that holds our picture of the universe together is stories, not facts.

Chapter 10

Why Dogs Sniff Each Other's Tails

I SOMETIMES BELIEVE that the motto of all dog lovers (and anybody who likes to tell tales) is "Never let the truth get in the way of a good story!" In the foothills of the Appalachian Mountains in southwest Pennsylvania, about midway between Pittsburgh and the state capital, Harrisburg, I found myself chatting with a man who bred Cairn terriers and had a sister who lived in Canada, who would eventually provide me with my own copy of this amusing little dog.

In addition to breeding dogs, Alex ran a boarding kennel and had just let several of his "guests" out into a large grassy exercise area. He walked to a nearby table which contained a pitcher of iced tea and poured me a glass. As we stood with glasses in hand, we watched the dogs doing their familiar greeting dance.

They milled around, each sniffing at each other's tail regions for a few seconds before offering invitations to run or play.

Alex leaned against the wood-slat fence, took a sip of his tea, then began to speak in a slow, deliberate tone that storytellers through the ages have used when starting to tell a tale.

"You know, I once asked my daddy why dogs sniff each other's tails, and he gave me a really nice explanation. It turns out that when God created dogs, He made them really smart. He gave them a secret language so that they could talk to each other and also gave them the gift of being able to understand human speech.

"Then he said to the dogs, 'You all have a job, and that is to look after humans, tend to their welfare, provide them comfort and help, and work with them when you can. While you are doing that, you are free to do anything else that you want to, except dance. Now I know that that must sound a bit odd, but you must understand that humans are both prideful and insecure. They believe that they are the smartest beasts on all the Earth. They feel that only they are capable of understanding language and speaking. They also believe that only they can make and appreciate music.'

"The dogs were all obviously puzzled by what God was saying, so He explained, 'I know that you all can understand human language, but since you speak only the language of dogs and can't actually speak the language of men, they will never know. I also know that you can appreciate music, but since you can't actually make music, humans don't know that either. However, if you begin to dance, then humans will begin to wonder whether you know other things, and they even might guess that you understand what they are saying too. Because they are insecure they will become suspicious and may wonder whether you have selfish personal motives for being around them. They may not let you stay near them any longer; they will not share their confidences with you or let you hear their plans. They may even drive you away from their homes because they think you capable of making evil, self-seeking plans that could harm them. All of this would damage your ability to be the companion and workmate of humans, which was the reason I created you. So you must never be seen dancing.'

"Now the dogs did like music and liked moving to rhythms, but they understood what the Lord was saying and agreed not to dance. For a long time everything went pretty well. Dogs listened to what people had to say and understood their plans and feelings so that they were better able to help out and provide comfort when that was needed. That is, until there was a very special human wedding party.

"There were only two villages on Earth at that time, and each kept pretty much to itself. However, a boy from one village met a girl from the other, and they fell in love and decided to get married. This meant that there would be a union between the two villages, and to celebrate this the people decided to have one blowout of a party. Since both villages would be involved there would be twice as many musicians available to make music—twice as many fiddles, twice as many banjos and guitars, two bass fiddles, two accordions, and two sets of drums. Never before had such a large, accomplished orchestra been assembled.

"All the people from both villages showed up to the party, and all the dogs came too, hanging around outside the village hall and waiting for the opportunity to grab a few scraps and leftovers. But as the dogs waited, they could clearly hear the music from that wondrous great band inside, The sounds were so happy and bouncy that the dogs began to sway and tap their feet.

"Suddenly a large handsome collie announced, 'God never really said that we couldn't dance; He just said that humans should never be allowed to see us dancing. Now the humans are all inside, and we're out here, so there will be no harm done if we allow ourselves the pleasure of a little dance.'

"With that he jumped up, stood on his hind legs, and began to twirl around. As he did so, his great bushy tail swept along the ground and a dust cloud began to form around him. One of the other dogs called out, 'Hey, Collie, you would dance a lot better if you weren't dragging that broom behind you.'

"The collie laughed. 'Easy enough to take care of,' he said, and he carefully unhooked his tail and tossed it on the ground, then continued to twirl and prance. The music was loud and the rhythm was catchy, so all the other dogs unhooked their tails and threw them onto a pile, with the collie's tail, and they also began to dance. Soon the whole pack of them were on their hind legs, spinning and bouncing to the music. A few dogs even began to yip and bark along with the happy tune that they danced to.

"I suppose that it was the noise that the dogs were making that was their undoing. Some of the people inside apparently

They threw their tails in a pile and began to dance.

heard that yapping and wondered what the dogs were doing. As they opened the big door of the village hall to check, the hinges squeaked, and light began to pour out into the yard.

"One of the dogs was alerted by this and shrieked an alarm. 'Stop! Quick, before they catch us dancing, grab your tails and act natural!'

"All the dogs dropped back onto their four feet and raced over to the pile of tails, each grabbing the first tail that they could, hooking it on and trying to act casual and nonchalant when the people looked out. One of the people said, 'I don't know what's going on out here, but I'll just stand guard and watch to make sure that nothing goes wrong and disturbs the other people at the party.'

"That person stayed out there until the party was over, at which time the people came out, gave their dogs some party leftovers, and then went home. In the morning the dogs woke up, feeling pretty pleased with themselves about having had a chance to do a little dancing and still having got away without being caught by their humans—at least until they had a good look at their tails in the morning light. Then they realized that every one of them had grabbed the wrong tail. They didn't know whose tails they actually had, but all of the dogs they had ever met had been there. So many dogs—so many tails! In the end the dogs understood that they might have to search far and wide to find their own true tails again.

"And that's why dogs sniff each other's tails: They are checking to see if they've found the dog that might have picked up their own God-given tail by mistake!"

The scientific explanation might not be as amusing as the story that Alex told me, but is still quite interesting. A dog's nose is tuned to a set of smells associated with pheromones, the odiferous chemicals secreted by animals that transmit information to other animals. The word derives from the Greek words *pherein,* meaning "to carry" and *horman,* meaning "to excite," because researchers originally thought that these smells told male animals when females were ready to mate. Today, however, we know that these personal chemicals carry a lot more information than just sexual readiness.

The source of pheromones is a special kind of sweat gland called "apocrine glands." High concentrations of these glands can be found around the genital and anal regions in all mammals (including humans). Pheromone smells contain more information about an individual than you find in the self-descriptions that people post on many social networking websites on the internet, such as Facebook or MySpace. They not only identify the sex, age, health, and mood of the individual, but also carry a lot of sexual information as well, such as where the female is in her menstrual cycle, if she is pregnant, or if she has recently given birth.

For dogs, reading pheromone scents is the equivalent of reading a written message about the status and feelings of another animal. Because of the location of the apocrine glands, many pheromone chemicals are found dissolved in dog's urine (and some are also carried in fecal matter), which means that a dog's urine can be "read" by other dogs. In effect, sniffing a fire hydrant or a tree along a route popular with other dogs becomes a means of keeping abreast of current events. That tree serves as a large canine tabloid containing the latest news items in the dog world. It may not contain installments of classic canine literature, but it certainly has a gossip column and the personals section of the classified ads. Dogs obtain an even clearer message by sniffing another dog in the area around its tail.

Many humans are embarrassed when a dog sniffs their crotch, but dogs do this for the same reason that they sniff each other—because that area is particularly rich in scents due to the concentration of apocrine sweat cells. Just as when they meet other dogs, strangers receive the most attention of this sort, especially if there is a tinge of sexual scent. People who have had sexual intercourse recently seem to attract this kind of attention from dogs, as do women in their menses, those who have recently ovulated, and those who have recently given birth (especially if they are still nursing their child). However, an inquisitive dog may poke his nose in the human groin just to learn more about a person.

Many humans have strong negative reactions when a dog starts examining their private body parts for scent messages. An interesting example of this was sent to me by the principal of an elementary school in Kansas City who wanted some advice. Apparently a parent complained that a boy in his school was sexually harassing her daughter. The instrument of his harassment was supposedly his golden retriever, which the boy allowed to "snoop, sniff, and snuffle" under the girl's skirts a number of times without trying to stop it. It turns out that each afternoon the eleven-year-old boy's mother would come by the school to take him home, often accompanied by the family dog who, when she opened the car door, would rush out onto the playground to greet his little master. Being a typically gregarious golden retriever, he would also take the opportunity to meet with as many of the children as he could. As might be expected, a quick crotch sniff was often part of this greeting process.

The school principal tried to reassure the complaining parent that none of the children seemed bothered by the sniffs and that the dog sniffed the groins of boys as well as girls. The parent responded angrily that "These children are too young and innocent to know that they are being sexually harassed," and that "The fact that other little boys are involved simply shows that homosexuality is part of his perversion." She demanded that the boy be expelled at the very least, and that perhaps the police or psychiatric experts be involved.

As I sat at my keyboard typing out a response with informa-

tion that the principal might use involving an explanation of
pheromones and apocrine glands, I doubted that this particular
parent would be satisfied and reassured by it. At the same time
it seemed even more unlikely that this parent would be any more
comforted by any interpretation that the big blond dog was simply searching for his own true God-given tail.

Chapter 11

The Laughing Dog

MUCH OF OUR bond with dogs comes from the fact that they are playful and uninhibited—as we could be if we weren't so worried about what other people think. We have actually bred playfulness into dogs by selecting for animals whose minds are much like those of wolf puppies all of their lives. It is their juvenile minds that makes dogs want to play and do those silly things that make us laugh. In humans the same behaviors would be evidence of a sense of humor.

Not all breeds are created equal, however, and some are more playful than others. Two animal behaviorists from the School of Veterinary Medicine at the University of California at Davis, Dr. Benjamin Hart and Dr. Lynnette Hart, had a group of experts rank the playfulness of fifty-six different breeds of dogs. "Playfulness" included a willingness to chase balls, Kongs, or Frisbees, and to engage in hide-and-seek and other games. Nobody would be too surprised to find that the Irish setter, English springer spaniel, Airedale, golden retriever, and poodle all ranked high in playfulness, whereas bloodhounds, bulldogs, and basset hounds ranked low. Here is a list of those playfulness rankings that the researchers produced for the breeds that they rated.

MOST PLAYFUL: Irish setter, English springer spaniel, miniature schnauzer, Cairn terrier, Airedale terrier, standard poodle, Shetland sheepdog, golden retriever, Australian shepherd, miniature poodle, German shorthaired pointer.

ABOVE AVERAGE: vizsla, fox terrier, Labrador retriever, Boston terrier, Yorkshire terrier, West Highland white terrier, toy poodle, German shepherd, silky terrier, Welsh corgi, shih tzu.

AVERAGE: dachshund, Weimaraner, Bichon Frise, cocker spaniel, Scottish terrier, dalmatian, boxer, pug, Maltese, beagle, collie, Brittany spaniel.

BELOW AVERAGE: Norwegian elkhound, Doberman pinscher, Chesapeake Bay retriever, Siberian husky, keeshond, Afghan hound, Pomeranian, Lhasa apso, Newfoundland, Old English sheepdog, Great Dane.

LEAST PLAYFUL: Samoyed, Chihuahua, rottweiler, Pekingese, Akita, Alaskan malamute, St. Bernard, basset hound, chowchow, bulldog, bloodhound.

As many owners will undoubtedly attest, playful dogs are not an unmitigated blessing. While such dogs are a joy to people who can handle the occasional bouts of chaos, they are an exasperation to those who cannot. Take the case of Joan and Flint. Joan is my wife, and Flint was her gift to me.

When I found myself without a dog for the first time in my life, Joan knew that I would go crazy if I didn't get one soon. She also knew that I had set my mind on a Cairn terrier because they are extremely playful and make me laugh. When you spend your whole life in a research laboratory or at a keyboard writing articles and books, having a dog that makes you laugh is better for stress reduction than a psychotherapist.

Having researched the breed, I hoped to find a Cairn terrier with some Melita blood in it because these are particularly sociable and friendly. Since the early 1930s, the Melita kennels on Vancouver Island in Canada had been run by Mrs. L. M. Wood, who really cared about the temperament of her dogs. Since she kept the dogs in group runs, any dogs that were aggressive or didn't get along well with the others were removed from her breeding program. Although Mrs. Wood had passed away a number of years earlier, the bloodlines were still clear enough to follow, and after a while I found a kennel with a good strong dose of the Melita lineage. Then Joan announced that she was giving me a puppy from their most recent litter as a Christmas gift.

I chose a puppy that was sociable and fearless. At that time he was only around three pounds of brown brindle fur. Named Flint, he would eventually become the number one Cairn terrier in obedience competition in Canada. He would also grow up to own a large part of my heart and to be the bane of my wife's existence. On this day, however, Joan smiled and held him, and we had a lovely family picture taken with the puppy under the Christmas tree.

The following year my wife bought me a twelve-gauge shotgun for Christmas. My daughter by marriage, Kari, assured me that there was a clear symbolic connection between the two Christmas gifts.

Flint was a constant trial for Joan, a prairie girl who had grown up in a family that kept large sporting dogs—mostly retrievers, pointers, and hounds. They were trained to pay attention, to do what they were told, and to work silently. These working dogs were allowed into the house only when they were to be fed or when the temperature dropped to something around minus 40 and everyone was feeling guilty about their welfare. Quiet, order, reliability, predictability, and unobtrusiveness are values that Joan cherishes in her own life and also demands from her dogs. A terrier as a house dog was something completely beyond her experience.

Joan had never encountered the likes of this dog before and was not amused by his favorite game, "The Barbarians are coming!" Flint played this game with great vigor at random intervals on carefully selected days and nights. It always began with him leaping into the air with a furious round of barking that was explosive enough to be heard throughout the entire house. Next he rushed to a door or window or leapt onto the highest surface he could reach, such as a bed or a sofa. He kept the cascade of noise going until someone in the family did the modern equivalent of grabbing bows and arrows and rushing to ramparts to defend the realm. His timing was such that the game always seemed to start when the house was quiet, because Joan was reading, sewing, or napping. Careful investigation would often reveal that the triggering event was usually something innocuous, like the wind brushing tree branches against the house.

To keep my wife from disemboweling Flint at such moments, I explained to her that terriers are specifically bred to bark. A working terrier absolutely must bark when it is the least bit excited or aroused because in earlier times this alerted the hunters to the location of the burrow where the dog had pursued an animal into its den. It was the sound of the barking underground that told the hunters where to dig to uncover the fox or badger. The earliest terriers, which were not so ready to bark, had to wear collars with bells on them to guide hunters in their chase and digging. Unfortunately, many terriers choked to death when their collars caught on some obstruction underground. Others died because the hunters could not hear the tinkle of bells when fox and terrier were lost under the ground in a final confrontation. A barking dog, however, could be heard, and hence found. My careful historical explanations were lost on Joan, especially when she had just been awakened two hours before the alarm clock was set to go off in the morning.

I tried to reduce the tension between my wife and my dog as best I could by pointing out the various clever and useful things that Flint did. I also tried to insert some humor into the situation, such as the day that I gave Flint a silly little doggy cap with the motto "Born to Bark" embroidered on it, to remind her of the heritage and behavioral predispositions of terriers. Joan was not amused.

Terriers were specifically bred to bark so that hunters could locate the burrow, and to help find the dog when he was in the den and underground.

"Why didn't you tell me all this before we got him?" she grumbled and then wandered off muttering about "nice quiet dogs, like golden retrievers and labs," and musing loudly to herself, "I suppose that he wants me to believe that there is a badger under the bed or something."

Flint had a mind of his own, and his likes and dislikes had no regard for Joan's preferred lifestyle. She would shoo him off a chair, only to see him immediately jump up on the sofa. She would push him off one side of the bed, only to have him jump back up on the other. She would scold him for barking at the door, only to have him jump up and begin barking at the window. One day she had some friends over for afternoon coffee. Flint hung around the group, demonstrating the sociability bred into him from his Melita heritage, also demonstrating the hopes of all dogs as he wandered around the room nosing at the visitors to test the possibility that one of them might scratch his ear or accidentally drop something edible. Joan became concerned that he might be annoying her guests, so she waved him away.

"Flint, stop bothering these people. Go find something interesting to do."

Flint's desire to follow her instruction was clearly aroused, and he dashed out of the room with a great sense of purpose. A few minutes later, he reappeared carrying one of Joan's undergarments. Evading capture, he proceeded to flagrantly snap it from side to side with great joy—to the amusement of the company and the dismay of my wife.

Flint and Joan almost reached an accord at one point, as a result of his mouse hunting. In the genes of every terrier is the ability and desire to eliminate rats and other vermin. People who have not had direct experience with this aspect of terrier behavior tend to think that the most efficient rat killers are cats. While cats are certainly efficient at killing mice, where stealth and patience are the most important qualities for the hunt, rats are often too large and vicious for cats to handle; hence, terriers were bred for the job. The general method that terriers use to dispatch their prey involves grasping the rat or other small mammal by the neck and giving it one or two swift shakes to break its neck. Even today, many farmers use terriers for rat control, especially in grain- or corn-growing regions. They often first send ferrets, or pump smoke or gas into areas where the rats are concentrated in order to rout them. Once rats are in the open, the terriers can rapidly dispatch the lot of them.

In our relatively rodent-free cities today, it is difficult to appreciate just how efficient terriers can be at rat killing. Back in the Victorian era, rat-fighting was a sport that was especially popular in lower-class regions of the city. It also drew a following from adolescents and young adults of the upper class. The young girl who later would become Victoria, Queen of England, enthusiastically wrote an account in her diary, describing the fun that she had had while attending one such spectacle. Terriers and rats were placed in a pit and were supposed to fight to the death. Side bets were often taken on the survival of dogs or rats, and other bets on the amount of time that some of the better dogs might take to finish off a particular group of rodents. A number of records survive describing some of the "superstars" of the sport. For instance, we know that one "champion" rat-fighter was "Tiny," a bull terrier, who weighed only 5 1/2 pounds. On one particular night he killed fifty rats (some of which were nearly as large as he was) in twenty-eight minutes and five seconds. His owner estimated that Tiny killed more than 5,000 rats during his lifetime, which would amount to a weight of around a ton and a half of rats!

Cairn terriers are no different from other terriers in that their desire to chase vermin, whether rats, mice, rabbits, or squirrels, is built in. The simplest way to arouse Flint to a frenzy of activity was to shine a spot of light on the floor with a flashlight and then to move the spot erratically around. A small moving target automatically elicits the pursuit response in terriers, and Flint would chase the spot with undying enthusiasm. The game would usually come to an end only when I grew too tired to continue, or the flashlight batteries began to fail.

One morning, Joan announced that "Flint has finally flipped out. Look at him."

As we turned to watch him, he made a mad dash across the room, stopped, and stared at something, then dashed in another direction. I looked more carefully at what he was doing and noticed that he was pursuing some glints of light on the floor. I tried to determine where these moving flashes were coming from and ultimately figured out that these glints were produced by the sunlight reflecting off of a large, cut crystal pendant that Joan was wearing around her neck. As these sparkling points of light randomly moved around the room, it turned on Flint's genetically-programmed hunting behaviors. When I pointed this out, Joan gave a theatrical sigh and then plaintively asked, "Do I have to now start taking the way that Flint's brain is wired into account before I get dressed in the morning?"

Flint, however, later proved to be quite an efficient "varminting" dog. In his life he would kill rats, mice, moles, gophers, and even an opossum. Most of this was done out at our little farm that I often use when I am writing or just hiding from the world, especially during the summer. Joan is really in charge of the farm, since I stay indoors to work while she putters around on her little tractor or works in her garden. Flint's thinning of the rodent population meant that her vegetable garden was more productive and there were fewer unwanted fur-bearing visitors trying to make their way into the house, garage, or cupboards where food was stored. His useful work was well appreciated by everyone in the house.

Sometimes Flint's spontaneous "hunt-and-kill" behaviors would lead to problems. Flint was one of the few dogs that I have owned that spontaneously watched television. You can often get dogs to watch televised images if you exert a bit of effort, such as lowering the TV set so that it is about at the dog's eye level and using images that are shot from a dog's point of view, say a foot or two from the ground. The images should have a lot of motion in order to attract the dogs to the location of the television and a carefully designed sound track containing lots of exciting sounds. Most dogs ignore those TV images that we enjoy, probably because they are high off the ground, don't contain the right kind of motion or content, or, most importantly, lack an attached scent. For these reasons televised images are seldom mistaken by dogs as real things.

Always scanning the environment for something that moved and might be chaseable, Flint was different. My spunky dog had first become interested when I turned the TV on to a program called *The Littlest Hobo,* a low-budget series that featured a German shepherd that wandered around the countryside, befriending various people and getting them out of trouble through his heroism and cleverness. Then, like an errant knight, at the end of each episode Hobo would wander away looking for his next adventure. When Flint saw this dog moving across the TV screen, his attention was immediately captured. He would stand up on his hind legs, the way that he often did at the windows to watch other dogs go by. If the dog disappeared from the screen, he would get closer and look slantwise in the direction that the dog had gone, perhaps trying to catch a glimpse of the disappearing furry star. After that he would always check the TV screen as he passed. If a dog or another animal was visible, he would stop to watch, and sometimes his tail would tremble as he studied the images, sometimes letting loose an excited whimper or a quick bark.

None of this caused any problems until the attack of the giant rats. I no longer remember the name of the film being shown on TV that night, but I do remember some of the content. It involved scenes in which rats were occupying some abandoned structure, or perhaps it was a tunnel or a mine. In any event

Flint's favorite television program was *The Littlest Hobo,* which starred a heroic German shepherd.

something happened and they became frightened and stampeded with a great shrilling sound, running toward the camera and past the hero of the film. The sequence began with some close-ups of the rats, which must have appeared to be truly giant rodents to Flint, since they filled the TV screen and, at his first sight of them, Flint froze. A low territorial growl started, and he began to quiver with excitement. When the rodent stampede occurred, with all of the accompanying frantic rat sounds, Flint could contain himself no longer. He launched himself off of the sofa and attacked the wooden stand on which the television stood. Growling, barking, slashing, chewing—he desperately tried to grab hold of the table leg and shake it to death. In a matter of moments the wooden leg of the TV stand looked like it had gone through a war. Meanwhile the rat scene had drawn to a close. The squeals were now gone and no rodents were visible any longer on the face of the tube. Flint backed off and looked up. He snorted once or twice through his nose, then with tail erect and legs stiff, proudly walked out of the room, pausing only once to glance at the TV to make sure that his job of saving us from the onslaught of vermin had been done well and was truly finished. I quickly rotated the TV table so that the damaged leg was against the far wall where the tooth marks would not be visible. I really didn't want to have to explain this new episode of genetically-generated terrier behavior to Joan, lest it further strain her relationship with my dog. Over the next couple of days I secretly repaired the damage and refinished the wood so that no evidence of Flint's rat-killing frenzy remained.

Flint's ability to hunt rodents did endear him to my wife for a while at least. In the city, we live in an old house built around 1916. It is not sealed as well as it should be, especially around the basement area. Every year, as the autumn rains fall and the weather starts to turn cold, mice work their way inside, starting in the basement and progressing up to the kitchen. With animals and grandchildren around, poison is not an option, and traps are inefficient.

Flint was of great assistance with our problem, hunting rodents the way that terriers were bred to do. With a degree of patience and dedication that would make cat owners envious, he turned into a fabulous biological mousetrap. Joan was quite pleased with Flint's proficiency. Typically, when he would kill a mouse he would leave it on the floor where it fell. When Joan noticed, she would gladly dispose of the small carcass and encourage Flint to keep up the good work. She would warmly praise him for his efforts, giving him a friendly pat and maybe even a treat.

Perhaps Flint saw this as his opportunity to make amends with that other human that he lived with, or perhaps he just reverted to being a terrier with a sense of humor. In any event, one morning Flint decided to make his peace offering to Joan. It was quite early, and Joan awakened to the gentle pressure of Flint's front paws resting on her. She looked down at him only to find that he had deposited a mouse on her chest—still warm, but quite dead. I fear that the gift was not accepted in the tender and accommodating spirit with which it was offered. Joan jumped up with a startled shriek and Flint began to dance happily around. He now knew that he had done something truly great and grand, since it was certainly causing such an interesting commotion on her side of the bed and such convulsive laughter on mine.

As with many terriers, Flint's motto in life was, "If two wrongs don't make a right, try three."

Chapter 12

The Sport of Queens

AT THE SHOUT of "tally-ho!"—the traditional British way of indicating that game has been spotted—two greyhounds are let loose. Looking like they were shot out of a cannon, they hurtle across the field. This is a sport called *coursing*, from the Latin *cursus* meaning "to run." It developed as a form of hunting, using dogs that hunt by sight and that have been specifically bred for their speed and ability to run down prey, as opposed to hunting dogs that use scent to track their quarry.

Perhaps the original sight hound was the greyhound, since we have pictures of what appear to be greyhounds being used for coursing by the Egyptian pharaohs. Their name comes, not from their color, but from a mistake in translating their early German name "greishund," which means "old" or "ancient" dog. The class of sight hounds includes Afghan hounds, basenjis, borzoi, Ibizan hounds, Irish wolfhounds, pharaoh hounds, Rhodesian ridgebacks, salukis, Scottish deerhounds, and whippets. All of these dogs have a characteristic physical form—large chests to provide space for big lungs that allow them to gulp the massive amounts of oxygen needed in a long chase, and narrow waists that give them the flexibility to bend so far that a single stride may move them more than a body length. Altogether they are built to be incredibly swift runners. Greyhounds are capable of running at close to 19 meters per second, or around 42 miles per hour.

While the original purpose of coursing was hunting, and virtually any animal could be coursed, the preferred targets were quite often hares, swift-moving animals that could give a sight hound quite a run, especially since the smaller hare could turn more quickly than the dogs and so had a real opportunity to escape. Aristocrats of the time were less interested in obtaining meat via coursing than they were in the sporting qualities of an interesting contest. Betting was part of this, with wagers on which dog would be first to reach the hare, which would catch it, and so forth. That this could be sport rather than subsistence hunting or pest control was recognized as early as 116 A.D. by the Greek historian Arrian, who wrote that "The true sportsman does not take out his dogs to destroy the hares, but for the sake of the course and the contest between the dogs and the hares, and is glad if the hare escapes."

The King of England, Henry VIII, set up coursing matches involving hares and sometimes deer as the quarry. He also implemented formal procedures to regulate betting. A well-bred aristocrat was expected to bet large sums of money without any apparent concern, and if he lost, he was supposed to accept that fate cheerfully. To the medieval mind, gambling was not a sin unless you cheated, and often failure to place or accept a wager was seen as evidence of indecisiveness or lack of courage. Henry was known to bet amounts that would be the equivalent of hundreds of thousands of dollars in a single day.

Like all of the royal Tudor line, Henry's daughter, Elizabeth I, loved the hunt, or at least loved to watch dogs coursing, and like her father she also loved the betting component of the sport. One afternoon at Cowdrey Park, Elizabeth watched a pack of greyhounds bring down sixteen deer. Displeased with this spectacle, in 1561 she ordered Thomas Mowbray, the Duke of Norfolk, to draw up The Law of the Leash to govern coursing events. Some have suggested that this was because she was appalled by the bloodletting; but that is not true. She was, however, distressed by the chaotic nature of the proceedings; the unfair advantage that the dogs seemed to have; the large number of dogs on the field, which made it hard to determine the actions of individuals; and the lack of a set of rules for determining excellence. All of these factors made it difficult and unsatisfying to place and resolve bets at such gatherings.

The Law of the Leash established that only two dogs would be coursed at a time. As the hare entered the running ground, the dogs had to be restrained until the hare had a head start of at least 80 yards. Each dog had to wear a distinctive color, and the judge followed the course and awarded points for speed, the ability to make the hare turn to avoid its pursuers, the first dog to knock the hare off its feet, and so forth. Winning was not necessarily dependent on catching the hare (although this did earn a high score), and often the hare escaped. Whether the hare was caught or not, a winner could be determined on the basis of point totals, and bets could be resolved. Because Elizabeth had ordered the drawing-up of the rules, coursing came to be known as "The Sport of Queens."

Even after the rules of coursing were established, it did not become a popular event with the general population for quite a while. In 1776 the Earl of Orford created the first coursing club open to the public, at Swaffham in Norfolk. But the sport did not reach national prominence until the founding of the Waterloo Cup in 1836. Considered the ultimate test for coursing dogs, the Waterloo Cup was held annually at the Altcar Club, near Liverpool. Named for the Waterloo Hotel in Liverpool, where the first promoters had met, the Cup was typical of most coursing events, which took place in a large fenced area so that spectators could watch the action. An assortment of escape holes large enough for a hare but not for a dog were built into the fences. Hares were captured and trained to know the escapes so that they would have a fair chance.

Eventually, coursing became incredibly popular in Britain. In the late 1800s daily crowds of 75,000 were not uncommon for the Waterloo Cup, and millions of pounds were bet on the outcomes. The winning dogs became national heroes, their names known by everyone. For example, the three-time Waterloo Cup winner Master M'grath was presented (by royal command) to Queen Victoria. When the competition was in progress, carrier pigeons were used to send the results to all major cities in Britain, and in London the Stock Exchange closed early when the news of the winners arrived. There was so much prestige associated with this event that simply being one of the sixty-four dogs nominated for entry was a matter of pride, and early advertisements for stud service or puppies included the line "Waterloo Cup nominator" to point this out.

Coursing came to the United States with the spread of farming to the wide grasslands of the West. Originally, coursing was simply part of an effort to control the populations of jack rabbits and coyotes, which were major pests for farmers and shepherds. Some of the earliest American Kennel Club-registered borzois were located in Kansas, and General George Armstrong Custer encountered Scottish deerhounds and greyhounds being used for coursing when he went to Texas after the Civil War. In the late 1800s, coursing changed from a hunting event to a competitive sporting event using live game and the same basic rules

The Waterloo Cup in the 1890s.

as in England. This sort of coursing, where sighthounds were let loose to chase live game in an enclosed area, was called "closed park coursing."

By today's more humane standards, a sport in which an animal is routinely set up to be killed and torn apart in public as an amusement does not fit our sense of morality. In many venues, closed-park coursing is illegal, but an alternative called *lure coursing* has much of the same competitive thrill as traditional coursing, and does not end in bloodshed.

In the late 1960s Lyle Gillette, a breeder of salukis and borzois, became interested in coursing. However, at that time some competitors would release their hounds too close to wire fences or other obstacles, and dogs were being injured. So he separated himself from those activities and set about trying to work out a system whereby sight hounds could be coursed more safely and the focus on killing the target animal eliminated.

Gillette's aim was to create a portable system that could be set up in a five-acre, open area and permit coursing with a lure that simulated the movements of live prey. After a lot of trial and error, he developed an apparatus in which a string was run through a set of pulleys planted in a field to form a course of 600 to 1,000 yards. Instead of a hare, the "prey" consisted of a piece of artificial fur or simply a plastic bag attached to the string. The arrangement of the pulleys allowed the path that the lure traveled to contain a number of quick right-angle turns, much like a real animal might make in attempting to elude pursuit. A motor provided the power, but the system also required the artistry of a lure operator to make it work. In the system of coursing that Gillette founded, which is the one still used today, the operator stands on a ladder, a platform, or the roof of a vehicle so that he can see the whole field and control the speed of the drag lure. The lure must be kept 10 to 30 yards in front of the lead dog— near enough to appear enticing and catchable while far enough away so that it won't be caught until it stops at the end of the run. Typically this involves moving the lure at speeds of between 35 and 45 miles per hour depending upon the speed and skill of the dogs involved.

Gillette demonstrated the prototype of his "lure coursing" system to the Mission Trail Borzoi Club in California, and they liked the idea of this new sport so much that they had representatives travel all over America and Canada to promote it. By 1972 there was enough support to form the American Sighthound Field Association, which established the formal rules, organized meets, and awarded titles.

Typically, dogs are run in pairs or sets of three and brought to the starting line wearing coursing blankets, which are bright pink, yellow, or blue jerseys. These allow the judges to easily identify participants. The dogs also wear slip leads, which are quick-release collars. The lure is started and, at the huntmaster's cry of "tally-ho!" the hounds are released to give chase. Two judges view the performance from different vantage points and award points for speed, agility, endurance, enthusiasm, and follow (pursuit of the lure). The absolute point score is not important; rather it is the dog's score in comparison to the other dogs in the competition that matters. Part of the dogs' reward at the end of the run is that they get to attack the lure.

The American Kennel Club only recognized lure coursing as a sanctioned event, and began holding matches and offering titles in 1991, by which time it was also becoming popular in Canada and Europe. Lure coursing was considerably slower to catch on in Britain than elsewhere, since coursing dogs with live hares as prey was still practiced there. The Waterloo Cup had been held annually for 156 years, despite the efforts of animal welfare advocates to stop it. The League Against Cruel Sports was established in 1924 to campaign against hare coursing, and several other animal welfare organizations made the Waterloo Cup a centerpiece of their campaigns.

It was not until 2004 that the British Parliament passed the Hunting Act, which outlawed fox hunting with dogs and also coursing events that used live hares. Some opponents argued that coursing with live animals was a venerable countryside tradition and that coursing competitions actually promoted the conservation of hares. In typical political fashion, the matter moved slowly, with elected officials unwilling to make a decision in the face of controversy. In 1999, after hearing from hundreds of citizens, visiting dozens of coursing sites, and expending hundreds of thousands of pounds, the commission reached a conclusion that could have been obtained from any randomly selected person they might have interviewed on the street. Their astonishing insight was that "We are similarly satisfied that being pursued, caught, and killed by dogs during coursing seriously compromises the welfare of the hare."

Chapter 13

Dogs That Wait and
Dogs That Come Home

THE FAITHFULNESS OF dogs to their masters is the stuff of which legends are made. Perhaps the most heartfelt of these stories are based upon the fact that dogs don't know, or at least don't seem to accept, the concept of death. For this reason there are histories from many places around the world of dogs that have kept a vigil, waiting patiently for loved masters who could not return because they had passed away.

Probably the most famous of these stories is that of Greyfriars Bobby. Bobby was a little Skye terrier who lived in Edinburgh, Scotland during the 1850s. Near the Greyfriars Church stood an inn, Traill's Coffee House, and every day at one o'clock,

Jock Gray, Bobby's master, would stop there for lunch. While he ate, Bobby happily gathered scraps from the floor and the occasional bone offered by a fond innkeeper. But the relationship had to end: In 1858 Gray died, and he was buried in the Greyfriars churchyard. Of course his close companion was there. The curator, James Brown, saw Bobby among the mourners when the grave was closed. And the following morning Bobby was found lying on the site, an action which was not permitted. Bobby had been allowed at the funeral as a gesture to his loving master, but dogs were not allowed in the churchyard. With regret, the curator chased him off. The next morning, however, he was there again, and again was shooed away. On a cold wet afternoon, a few days later, Brown again found the dog lying on the grave, shivering with cold and refusing to move. At that same moment, the innkeeper arrived, having followed Bobby from the inn where he had appeared each afternoon at around one o'clock to beg for scraps of food. Each time he finished and drank some water, the terrier would leave in a purposeful manner, making the innkeeper curious. Both men agreed that this devotion had earned Bobby the right to make the churchyard his home. They erected a small shelter nearby.

People soon learned of Bobby's devotion, and some time later, when the question of Bobby's dog license arose, the Lord Provost of the day paid it personally. For fourteen years, until his own death in 1872, this faithful dog kept constant watch, waiting for his master to return. Bobby now rests in the churchyard not far from Jock Gray. The Baroness Burdett-Coutts was so moved by Bobby's long vigil that she erected a statue of the little dog just outside the gate, and it was unveiled without formal ceremony, on November 15, 1873. The inn that Bobby and his master frequented is still there, but is now called Greyfriars Bobby Inn.

From the other side of the world comes a remarkably similar story. Hachiko was an Akita owned by Dr. Eisaburo Ueno, a professor at Tokyo University. Every day Hachiko would accompany his master to the train station to see him off. Using the internal clock that many dogs have, he would return to the station each afternoon at three o'clock to greet his master. On May 21, 1925, Professor Ueno did not return; he had died that afternoon at the university. Hachiko, the loyal Akita, waited at the station until midnight. The next day, and for the next ten years, Hachiko returned to the station and waited for his beloved master before walking home alone. Nothing, and no one, could discourage Hachiko from maintaining his nightly vigil. As in Bobby's case, people were moved by this loyalty and fed and watered the dog. Until he himself died in March 1934, Hachiko never failed to appear in his place at the railroad station. A statue of the Akita was erected near the Shibuya Station. It was destroyed during the war, and replaced in 1948, re-created exactly by the son of the artist who had produced the original statue. This train station is now a place where lovers go to touch the statue of the dog and to pledge that they will be as faithful to each other as Hachiko was to his master.

An American story of a dog who waits takes place in Montana. In the late summer of 1936, a man died and his remains were put in the funeral car of the Great Northern Railway in the presence of only one mourner, a collie named Shep, who tried to get into the car by scratching at the door. As the train started moving eastward down the track, the collie whimpered plaintively. Then Shep set up a watch for his master's expected return, digging a shelter under the depot and waiting for incoming trains. Even in the harshness of Montana winters, he met each train, and checked all passengers with a glance, a sniff, and a cursory wag of the tail, often seeming disappointed when "his person" was not there. A train conductor gave the story to a newspaper, and soon people were sending gifts of food and money to care for the dog. For over five years Shep watched and waited, gradually becoming stiff and slow moving. On January

Hachiko waits for his master's return.

12, 1942, he stood on the tracks to watch a train arrive from the east and, his eyes dimmed and his old legs too stiff to move out of the way, he was hit and killed. Some railway men built a concrete statue of him, which the Great Northern kept lit up at night for many years as a tribute to this loyal dog.

Unlike Bobby and Hachiko, however, whose statues have become tourist sites, Shep's statue is not the final legacy of his faithfulness. The money that was sent by caring people to support the collie's vigil actually grew to a sizeable amount. After his death the funds were used as an endowment to the Montana School for Deaf & Blind Children. Even today, the "Shep Fund" is still providing children with hearing aids, guide dogs, and educational tuition assistance. Dogs may not know about death, but they may know that loyalty, faith, and love are blessings and transcend the life of any individual.

Perhaps the extreme opposite of the dogs that wait are the dogs that are separated from their families and do everything in their power to return to them. Sometimes these events result in astonishing journeys by the dogs. A question that often prompts heated discussions is whether dogs have a special sense that allows them to find their way back to their loved ones from wherever they have been left or lost. Some people feel that dogs have a kind of homing instinct like that of pigeons, that can find their way back to their own lofts over distances of many miles. Other people feel that this homing talent is a myth that started with the film *Lassie*.

The first Lassie film was based on one of the most famous and best loved of all dog stories, *Lassie Come Home*. In Eric Knight's tale, the beautiful collie, Lassie, escapes from the Duke of Rudling's harsh kennel keeper to find her way back to the home where she was born. Lassie's journey takes many weeks, during which she travels from Scotland to Yorkshire, a distance of about 400 miles. While Lassie's story was fictional, Knight actu-

Some dogs seem to have an astonishing sense of direction and location.

ally drew his inspiration from a newspaper article about a real dog that found its way home over a distance of several hundred miles.

There are many such true stories which lend credence to certain dogs' ability to sense direction. Consider, for instance, the story of Marin, a Shetland sheepdog that looked for all the world like a miniature version of Lassie in sable and white. Marin was owned by Rita Johnson, who lived in a house on the northern edge of Los Angeles. In the spring of 1991, Rita went to visit some friends in San Francisco and took Marin with her for the holiday. The traffic was heavy, and Rita was running a bit late. Fearing that her friends might become worried, she stopped about 20 miles south of San Francisco to phone ahead to say she was still en route. While Rita was on the phone, her car was stolen—with Marin in it. The car was later recovered in San Francisco, but Marin was nowhere to be found. Rita was heartbroken as she returned home without her dog. Late that fall, nearly five months after she had lost Marin, Rita heard some whimpering outside her door. When she opened it, there was a thin, gray, shaggy dog, with sore, bloody feet. Out of compassion she took the dog in, and gave it a bath only to find that when the gray washed away, the familiar sable and white of Marin appeared. This small sheepdog had traveled nearly 60 miles farther than the fictional Lassie, to find his way home.

Of all of the dogs that have returned after being lost, the distance record is probably held by a fox terrier named Whisky. In October 1973, his owner, a truck driver named Geoff Hancock, stopped for a cup of coffee at a cafe near Darwin, Australia. Although he was gone for only a short time, it was enough for Whisky to jump out of the cab and get himself lost. Amazingly, Whisky eventually showed up again at Hancock's home in Melbourne. This little dog had traveled an average of 7 miles a day, for nine months, a total journey of nearly 1,800 miles!

Then there is the story of Neptune, a dog that tracked his

owners for over 50 miles. Compared to the journeys of Marin and Whisky, this might not seem like much of a distance, but this dog *swam* all the way. Neptune was a Newfoundland dog, which is a powerful breed with strong swimming abilities. While the dog was on board a boat being towed down the Mississippi to New Orleans, the craft hit some obstruction that caused it to lurch violently and tossed Neptune overboard. The owner of the boat couldn't stop, since it was being towed along with some barges by a distant tugboat. Thus Neptune's family could only watch as Neptune fell farther behind and disappeared in the distance. Rather than immediately swimming ashore, however, Neptune simply kept paddling downstream, following the boat that contained the people that he loved. Three days later, much to the amazement of his family, Neptune appeared, swimming alongside their craft, having tracked them to their mooring in New Orleans. The great black dog casually jumped on board as if nothing unusual had happened.

Sometimes dogs benefit from a little assistance in finding their way back to their owners. Margaret Wilson, for instance, lives in downtown Dallas, Texas, and was returning home by car from a visit to her sister in Pittsburgh, Pennsylvania, accompanied by her big retriever, Dash. Wilson would take Dash on long trips both for companionship and for security, because she felt a little vulnerable as a woman traveling alone. About halfway home she stopped at a motel not far from a town called Litchfield, Illinois. It was a warm night and the air conditioner in her room was noisy, so she turned it off and left her windows open. When she woke up in the morning, she discovered that Dash was missing. She guessed that he had jumped out of one of the low-set open windows during the night.

Margaret was in a panic. She was far from home, in a rural area, and Dash, being a city dog, probably would not be able to fend for himself in these conditions. He certainly would not have a clue about how to find her in this unfamiliar place. She got into her car and drove several miles in both directions looking for him, and calling his name, without any success. Finally, she returned to the motel to speak to the manager about who she might call to help her find her dog. The manager called the state police. One of their patrol cars happened to be nearby, and the state trooper was there in a few minutes. He listened to Margaret's tale of woe, then made a suggestion.

"Look," he said, "if your dog is a city dog, it is unlikely that he will wander too far from the road, since paved areas and vehicles are most familiar to him. This motel is the closest thing that your dog has to a recognizable place, so if you can stay over for a couple of nights, I have a plan that might work. What your dog needs is some kind of 'signpost' that will act as a sort of map to bring him back to you. Since your dog can't read posters, something with your scent will work best. Just leave items with your scent along the side of the road and if your dog finds one of them there is a chance that he will follow the trail back here to you."

Margaret was not sure that this would work, but since she had no other plan she decided to give it a try. Her suitcase was full of dirty clothes from her trip, so, despite feeling extremely silly, she placed items of her apparel every 50 yards or so along the road, for a couple of miles in either direction. Then she sat in front of her motel room to wait. Nothing happened all day, and in the evening she sadly went to bed.

In the morning Margaret got up, hoping that Dash had returned, but he had not. So she committed herself to at least one more day of waiting. As twilight approached, she left her post in front of her room to get something to eat at a little diner across the road. As she glanced down the road to check for traffic she saw the familiar silhouette of a dog that appeared to be sniffing at one of the pieces of clothing items she had laid by the roadside.

Margaret reports, "I yelled, 'Dash, come here!' and he lifted his head and charged over to me, carrying a piece of my underwear in his mouth."

That night Margaret slept with the windows closed, and in the morning put Dash in her car and started down the highway to collect her clothing.

"I don't know who, or why," she said, "but sometime before we were ready to leave in the morning, someone traveled up and down the road and picked up every piece of clothing that I had dropped. I couldn't even find a single sock! I went home with an empty suitcase—but I did have my dog!"

Obviously dogs have gone missing and never returned in spite of attempts to find them or help them to find their owners. Because of this we can certainly conclude that dogs' homing sense is not perfect. Take the case of Elvis, a mixed breed that looked much like an Australian cattle dog. He was owned by Ed Wood, a fashion salesman who lived in Philadelphia but traveled widely through the Eastern states selling his line of clothing. Sometimes on shorter sales trips, Ed would take Elvis along for company. On one trip he stopped at a motel in Harrisburg, Pennsylvania.

Dash comes home carrying a "clue."

Ed believes that he probably had not fully closed a sliding door when he decided to take a nap; when he awoke, Elvis was gone. A search of the surrounding area failed to find him, and Ed sadly returned home alone. One morning six weeks later, however, Ed was getting dressed, and the television was on. A woman from a local animal shelter was displaying the "Dog of the Day," usually a stray that had been turned in to the shelter and was being offered for adoption. She announced, "All that we know about this dog is that he was wearing a tag which said that his name is 'Elvis.' " Ed stared at the screen and then rushed off to the shelter to collect his long-lost pet. Had Elvis spent the last six weeks wending his way east? Had Elvis displayed a Lassie-like talent in making the 102-mile journey from Harrisburg to Philadelphia? Not exactly. You see, Ed was on another of his sales trips. This time he was in a motel in Pittsburgh. It was there that he had the good fortune to see his dog on television. Elvis actually had been traveling west, and had covered a total of 189 miles—in the wrong direction!

Apparently, not every dog is a potential Lassie.

Chapter 14

Can a Dog Really Love?

IF YOU WANT to cause a commotion in a psychology department or any other place where animal and human behavior is studied, all that you have to do is to claim that your dog "loves" you. Skeptics and critics, as well as some ardent supporters, will crowd the halls to argue the pros and cons of that statement.

Among the skeptics you will find the veterinarian Fred Metzger of Pennsylvania State University, who claims, "Dogs probably don't feel love in the typical way humans do. Dogs make *investments* in human beings because it works for them. They have something to gain from putting so-called emotions out there." Metzger believes that dogs "love" us only as long as we continue to reward their behaviors with treats and attention.

Most dog owners, however, know that dogs *can* truly love

people. Take the story of Rocky and Rita from the Finger Lakes region of New York State, near Rochester. Rocky was a solid 65-pound boxer, classically colored with a chestnut brown coat and a white blaze on his chest. At the time of this story, Rocky was three years old and Rita was his eleven-year-old human companion. Rocky had been given to Rita when he was ten weeks old, and she immediately bonded with him, petting him, hand-feeding him, teaching him basic commands, and letting him sleep on her bed. She thought that he was the most handsome dog in the world, and on rainy afternoons she would place him in various poses wearing hats, ties, or scarves, then sit and sketch pictures of him. When she was not in school, the two were always together and within touching distance. The family would often fondly refer to the pair as "R and R."

Rita was relatively timid and shy, and as Rocky grew, he brought her a sense of security. When Rocky was next to her, Rita felt confident enough to meet new people and go to unfamiliar places. Rocky was not only her friend and confidant, but also her defender.

When encountering strangers, Rocky would often deliberately stand in front of Rita, as a sort of protective barrier. He seemed fearless. Once when Rita was about to enter a store, two large men dressed in biker outfits burst out of the door, yelling at the shopkeeper and nearly knocking Rita over. Rocky rushed forward, putting himself between the frightened girl and the two threatening men. He braced himself and gave a low rumbling growl that carried such menace that the men backed off and gave the child and her guardian a wide berth.

There was, however, one flaw in Rocky's armor—a fear of water that was so extreme that it was almost pathological. Boxers are not strong swimmers, and are often shy of the water, but Rocky's fears stemmed from his puppyhood, when, at the age of seven weeks, he was sold to a family with an adolescent boy who had emotional problems. He acted as if the attention that the new puppy was getting somehow meant that he was less important. In a jealous rage, he put the puppy in a pillow case, knotted the top and threw it into a lake. Fortunately, the boy's father saw the incident and managed to retrieve the terrified puppy before he drowned. He scolded the boy and returned to the house. The next day the horrified parent saw his son standing waist-deep in the lake again trying to drown the struggling puppy by holding him underwater. This time Rocky was rescued and returned to the breeder for the dog's own safety.

These early traumas made water the only thing that Rocky truly feared. When he came close to a body of water, he would pull back, distressed. When Rita would go swimming in the lake he would pace along the shore trembling and whimpering. He would watch her intently and would not relax until she returned to dry land.

One late afternoon, Rita's mother took R and R to an upscale shopping area located along the edge of a lake. It featured a short wooden boardwalk built along the shore over a sharp embankment that was 20 or 30 feet above the surface of the water. Rita was clomping along the boardwalk, enjoying the way the sounds of her footsteps were amplified by the wooden structure when a boy on a bicycle skidded on the damp wooden surface, hitting Rita at an angle that propelled her through the open section below the lowest guard rail. She let out a shriek of pain and fear as she hurtled outward and down, hit the water facedown, then floated there unmoving. Rita's mother was at the entrance of a store 100 feet away. She rushed to the railing, shouting for help. Rocky was already there, looking at the water, trembling in fear, and making sounds that seemed to be a combination of barks, whimpers, and yelps all rolled into one.

We can never know what went through that dog's mind as he stood looking at the water—the one thing that truly terrified him and that had nearly taken his life twice. Now here was a frightening body of water that seemed about to harm his little mistress. Whatever he was thinking, his love for Rita seemed to overpower his fear, and Rocky crawled through the same open space in the rail and leapt forward and into the water.

One can thank the genetic programming that allowed the dog to swim without any prior practice, and he immediately went to Rita and grabbed her by a shoulder strap on her dress. This caused her to roll over so that her face was out of the water, and she gagged and coughed. Despite her dazed state, she reached out and managed to cinch her hand in Rocky's collar, while the dog struggled to swim toward the shore. Fortunately the water was calm and they were not far from shore. In a few moments Rocky reached a depth where his feet were on solid ground. With great effort he dragged Rita until her head was completely out of the water, then stood beside her, licking her face, while he continued to tremble and whine. It would be several minutes before human rescuers made it down the steep rocky embankment, and had it not been for Rocky, they surely would have arrived too late.

Rita and her family believe that it was only the big dog's love of the little girl that caused him to take what he must have considered a life-threatening action. This clearly casts doubt on Dr. Metzger's theory that dogs don't love us but act only out of self-interest. Why should Rocky behave in a way that he certainly felt would threaten his life? Surely, if he was evaluating the costs and benefits of his actions, he would have known that, even in Rita's absence, the rest of the family would be around to feed him and take care of his needs.

Marc Bekoff, a behavioral biologist at the University of Colorado, has a different interpretation. He points out that dogs are social animals. All social animals need emotions, in part as a means of communication—for instance, you need to know to back off if another animal is signaling that he is angry by growling. More importantly, however, emotions keep the social group together and motivate individuals to protect and support each other. Bekoff concludes that "strong emotion is one of the foundations of social behavior and the basis of the connection between individuals in any social group, whether a pack, a family, or just a couple in love."

Recent research has even identified some of the chemicals associated with feelings of love in humans. These include hormones such as *oxytocin*, which seems to help people form emotional bonds with each other. One of the triggers that cause oxytocin to be released is gentle physical touching, such as stroking. Dogs also produce oxytocin, and one of our common ways of interacting with dogs is to gently pet them, an action that probably releases this hormone associated with bonding. If dogs as social animals have an evolutionary need for close emotional ties, and have the chemical mechanisms associated with loving, it makes sense to assume that they are capable of love, as we are.

Rocky's fear of the water was absolute, and never did abate. He continued to avoid it for the rest of his life, and no one ever saw him so much as place a foot in the lake again. No one, at least not Rita or her family, ever doubted his love for her. He lived long enough to see Rita graduate from high school, at which time she posed for a photo in her cap and gown. Beside her sat a now much older boxer. The smiling girl had an arm around the dog, and her hand was cinched in his collar, as it had been the day that Rocky unambiguously showed Rita just how much he loved her.

Talking with Dogs

ODIN

Chapter 15

What's in a Name?

WHEN YOU HEAR the names Lassie, Rover, and Fido, you immediately know that we are talking about dogs. A dog's name is perhaps the single most important word that he will ever learn. Think of it this way: a dog lives in a sea of human sounds and, with the language ability only of a human two-year-old, it has to decide which words are directed at it and which are not. Thus if you say to another family member, "I am going to come over and sit down and watch some TV," how does the dog know whether the words "come," "sit," and "down" are not meant as a command for him? Obviously, if you look directly into the dog's eyes and have his full attention, the "sit" or "down" clearly are directed at him, and he should know that you mean for him to respond. In the absence of such body language, however, the dog's name becomes the key to his understanding. In effect, a dog's name becomes a signal that tells him that the next sounds that his master makes will have some impact on his life. The dog's name is an attention-getter that translates into something like, "The next message is for you."

This means that you should be precise when talking to your dog. Each time you want your dog to do something, start off by saying his name. This means that "Rover, sit" is proper dog talk. On the other hand, "Sit, Rover," is not good grammar for a dog, since the command that you want the dog to respond to will have disappeared into the void, before he has been alerted that the noises that you are making with your mouth are addressed to him. The result is that when you say "Sit, Rover," and nothing meaningful follows his name, you may end up with a dog simply staring at you with that "Okay-now-that-you-have-my-attention-what-do-you-want-me-to-do?" look that we all have seen so many times.

All of my dogs have three names. The first is their official name, which is the name that is registered with the kennel club and appears on their pedigree certificate. These are usually marvelously pompous or meaningless, such as "Remasia Vindebon of Torwood," "Rashdyn's Braveheart Rennick," or "Solar Optics from Creekwood." The American Kennel Club gives you thirty letters with which to come up with this formal title (including spaces, apostrophes, and hyphens). If you decide on a name that somebody has already used, then some of those letter spaces must be devoted to a number to distinguish your dog's name from all others. (I sometimes wonder whether there is a collie out there with the name, "Lassie, number 6,654,521.") You also can't use any word that might be similar to an earned kennel club title such as Champion or Champ, and of course, no obscenities are allowed.

Some dog breeders decide on the dog's registered name before a prospective buyer even arrives, but most breeders want the dog's new owner to have some input. Some breeders adopt a sort of "theme" for each litter. The simplest is alphabetical. Thus a breeder might require new owners to give all of the dogs in their first litter names that start with the letter "A," and their next litter, names with the letter "B," and so forth. This helps breeders sort out which dogs came from which litters. For example, my beautiful black flat-coated retriever was in an "O" litter, so I named him "Odin," after the Norse god who walked the plains of Valhalla accompanied by his two great black dogs. Other breeders might use more exotic themes, such as names of rivers, cities, or in the case of my beagle's breeder, movie stars of the 1940s. Thus my beagle became OSheehan's Call Me John Wayne.

The dog's second name is his "call name," which is the one that we actually use when we talk to him. After all, you really don't want to be standing out in your backyard yelling "Remasia Vindebon of Torwood, come!" For instance the call name of OSheehan's Call Me John Wayne is "Darby." My dogs typically get names like Wizard or Odin. Over the years I have found that two-syllable names seem to roll off my tongue more easily and produce a better response. It is also easier to include an emotional tone when needed—either inviting or demanding— with a two-syllable name. Usually people like there to be some link, no matter how tenuous, between their dog's call name and their registered name. Thus "Koy's Abracadabra Alchemist" got the call name "Magic," while "Remasia Vindebon of Torwood" simply became "Vinny."

Personally, I do not like to use human names for my dogs since it can be confusing. I remember being at an obedience ring in a dog show with a friend named Emma. As we sat there she told me that she kept hearing someone calling her name, but she couldn't recognize who was doing it, and it was driving her to distraction. At just that moment, a woman with a springer spaniel arrived at the ring, looked at her dog, and said "Emma, sit!"

and we had our answer. It is especially embarrassing if you name your dog and then later remember that you have a family member with the same name. Uncle Freddy might appreciate it if you named your child after him but might feel somewhat put out if you name your Jack Russell terrier "Freddy."

It is also a good idea to avoid names that are too cute. The comedy of calling your dog Fur Face or Butt Head will wear quite thin over the months and years that your dog will be with you. Furthermore, think about whether you really want to be heard loudly calling your dog in a public place when your dog is named "Fuzzykins" or "Noodle."

As of this writing, the twenty most popular call names for male dogs in the United States are:

1.	Max	11.	Charlie
2.	Jake	12.	Jack
3.	Buddy	13.	Harley
4.	Bailey	14.	Rusty
5.	Sam	15.	Toby
6.	Rocky	16.	Murphy
7.	Buster	17.	Shelby
8.	Casey	18.	Sparky
9.	Cody	19.	Barney
10.	Duke	20.	Winston

while the most popular call names for female dogs are:

1.	Maggie	11.	Dakota
2.	Molly	12.	Katie
3.	Lady	13.	Annie
4.	Sadie	14.	Chelsea
5.	Lucy	15.	Princess
6.	Daisy	16.	Missy
7.	Ginger	17.	Sophie
8.	Abby	18.	Bo
9.	Sasha	19.	Coco
10.	Sandy	20.	Tash

I must admit that I was surprised to find that neither the names Lassie or Snoopy (the dog from the popular *Peanuts* comic strip by Charles Schulz) appeared in these lists of popular dog names. I was even more surprised to find that both appeared in the lists of most popular names for cats!

All of my dogs also have a third name, a group name, which for me is "Puppy." This is their alternate name. Thus when I yell, "Puppies, come," I expect all of my dogs within earshot to appear at a run. A friend who only has male dogs uses the word, "Gentlemen," while another (a former officer in the Army Tank Corps) uses the group name "Troops."

The kennel club allows only one change of the dog's registered name during his lifetime. Dogs, however, are more flexible, as long as there is a chance for them to learn their new label. My daughter by marriage, Karen, for example, had a dog named Tessa, which became Tessa Bear, then for many years was simply Bear, and eventually went back to being Tessa. Tessa handled all of these changes of title with equanimity.

Tessa's name-change problems were minor compared with those of some dogs. For example, the Skye terrier owned by Robert Louis Stevenson, best known for writing classics such as *Treasure Island* and *The Strange Case of Dr. Jekyll and Mr. Hyde*, was initially named "Woggs," which was then changed to "Walter," which was then modified into "Watty," transformed to "Woggy," and finally changed to "Bogue."

Dogs' call names are changed for many reasons. Sometimes this just happens because of family interactions. So when my brother Arthur was very young he could not say the word "dog" clearly and it came out as "glock." When he wanted to call my boxer, Penny, he would just stand there and yell "glock." My family thought that was amusing, and after a while everyone was calling Penny "Glock." She had no problems with the name change as long as a treat was available when she responded to her new moniker.

Adopted dogs often undergo name changes as well. I adopted an older Cavalier King Charles spaniel after his owner suffered some medical problems and couldn't keep all of her dogs. A handsome animal, he came with the name "Banshee." I tried to use that name for him when I got home, but my professional soul had great difficulty calling a male Scottish dog using the name of a female Irish ghost. I did ask Katie why she had named him Banshee, and she had a perfectly sensible reason. When he was very young, he used to make a sort of "woo-woo" sound, she said. When she lived in England and the wind made a similar sound as it blew past the roof of her house, her family referred to it as a "banshee wind." She never bothered to look into the origin of that term, and it just seemed appropriate for this noisy pup. However, I never got used to it, and ultimately renamed him "Bam Bam" after the little boy in the cartoon series *The Flintstones*. I felt that the first syllable of his old and new names sounded sufficiently similar and was apparently correct because he responded to his new name the very first time that I used it, and has been happily responding to it ever since.

All of this simply demonstrates that any sound that is consistently used to address a dog can come to be its name, at least for a while. I had an interesting experience with one dog, a Siberian husky named Polar. I had been invited to be a special guest speaker at a scientific conference which was being held at a ski resort, and I was housed in a cabin that I shared with Paul, one of the conference program directors. Paul lived within driving distance of the resort and had brought Polar with him. He knew that I liked having dogs around me at all times and thought that it might help me get over the separation pangs that I have when I am on the road and away from my own puppies.

It was interesting watching Polar and Paul interact. Although Paul clearly loved the dog, he was having some trouble controlling this rambunctious, bouncing ball of fur. As soon as the car door opened, Polar dashed out. Paul yelled "No," and the dog

It seemed inappropriate to name a male Scottish dog after a female Irish ghost.

obediently came back to his side. When I went to greet the dog, it jumped up on me and Paul again brought it back down to the floor and to his side with a sharp "No!" That evening, as Paul and I sat chatting over a drink, Polar began nuzzling him to try to get one of the pretzels we had in the bowl between us. Again a quick "No," and Polar settled down with a sigh. Later that night there was some commotion on Paul's side of the room. Polar had tried to snuggle his way onto Paul's bed and was pushed off with a sharp "No." In the morning the first sounds that I heard were Paul telling Polar, "No, it's too early. I don't want to get up yet." Then a few minutes later, "No, let me sleep. I'll let you out in while."

Later, over dinner, Paul confided that he sometimes felt that he didn't have the dog under control much of the time: "For example, there are times when I don't even think that Polar knows his own name."

"Polar knows his name," I told him, "however, you might not." In response to his puzzled look I said, "We'll run a little experiment when we get back to the cabin tonight."

Later that evening we returned to the cabin. Once there, I instructed Paul to stand in the kitchen area, and I took Polar with me out on the deck just off the bedroom. I was petting Polar, who seemed contented at the attention that he was receiving, when Paul, standing in the kitchen shouted (as we had prearranged), "No!" Polar stood up and quite obediently trotted off to his master. On the basis of what he had experienced during his life, the sound that he had heard most frequently associated with personal consequences for him was "No." In Polar's mind, "No!" was his name!

Chapter 16

The Universal Dog Language Translator

WHEN I WAS six or seven years old, I used to dream about King Solomon's ring. The traditional stories say that it was a silver ring that bore the king's personal seal and the true name of God and that it gave him the ability to understand and to speak with animals. When Solomon died, that ring was hidden in "a great house with many doors." I dreamed of finding that house and the right door.

Why I needed Solomon's ring became clear to me every Sunday evening. That's when I would sit on the living room floor in front of the big family radio with my beagle, Skippy, and listen to a radio adventure show featuring my favorite movie star. I would wait for the beginning strains of the show's theme, the folk song "Greensleeves," and above the music, I could hear her voice growing louder and closer as she barked a happy greeting.

Of all of the dogs that starred in movies, Lassie is the most familiar. Perhaps Lassie's most unusual role, however, was in a radio series which ran from 1947 through 1950. Given the media mentality of today, producers of a contemporary radio series involving a dog might argue that it would be necessary to give Lassie a human voice so that we could hear her thoughts and know what she wanted to say; however, these early radio episodes were true to the character of Lassie's screen persona, and she communicated only by barking.

A typical episode might have Lassie running into the kitchen barking frantically. The mother of her family asks "What's wrong, girl?" and Lassie barks and whimpers excitedly.

As the interpretation becomes clear, Mom says in a worried voice "Oh no—Timmy has fallen down the well! You go get Doctor Connors. I saw him stopping by the Johnson place just a little bit down the road. Meanwhile, I'll go see what I can do."

The woman runs out of the house toward the well, while Lassie barks and races off for help. The doctor will of course understand every bark and whine that Lassie produces and will also know to bring Mr. Johnson and a rope when they come to the rescue.

I really felt jealous of the ability of Lassie's family and neighbors to understand the language of dogs. As I fondled Skippy's soft long ears, I wondered why I was so linguistically inept. I promised myself that one of my tasks in life would be to learn to interpret dogs. This was one of the motivations which ultimately led me to become a behavioral researcher.

Science has made it clear to us that when it comes to communication based on sounds, there appears to be a universal code used by most animals. It is based on three aspects of the noises that animals make, namely: pitch, duration, and frequency (or repetition rate).

Pitch refers to whether sounds are high and squeaky or low and grumbly. Low-pitch sounds, such as a dog's growl, usually indicate threats, anger, and the possibility of aggression. These are interpreted as meaning, "Stay away from me." High-pitch sounds mean the opposite, with the animal asking to be allowed to come closer or suggesting it is safe to approach.

When it comes to sound *duration*, generally speaking, the longer the sound, the more likely the dog is making a conscious decision about the nature of the signal and subsequent behavior. Thus a threatening growl from a dog that intends to hold his ground and not back down will be low-pitched and last several seconds or longer. In contrast, a growl coming in shorter bursts and only briefly held indicates an element of fear from a dog worried about whether it can successfully deal with an attack.

The final dimension is *frequency* or repetition rate. Sounds repeated often and at a fast rate indicate a degree of excitement and urgency, whereas sounds that are spaced out or not repeated at all usually indicate a minimal level of excitement. So an occasional bark or two at the window is only an expression of mild interest. A dog barking in multiple bursts and repeating this many times within a minute, on the other hand, feels that the situation is important and perhaps even a potential crisis.

With only three major sound dimensions, (and perhaps a few minor ones, like loudness) Matsumi Suzuki of the Japan Acoustics Laboratory realized that it might be possible to program a sort of "canine voice recognition" machine which could be used to interpret dog language, or at least the emotional content of dog communication. The result was a little device called the "Bow-Lingual," which has a transmitter that clips on to the dog's collar and a receiver that analyzes the noises that the dog makes and displays an interpretation for you to read.

Actually Bow-Lingual's processing divides the dog's utter-

A sound sensor and transmitter is worn on the collar.

ances into six emotional categories: happiness, sadness, frustration, "on-guard," assertiveness, and neediness. It then tags the analysis with one of 200 preprogrammed phrases. The particular "translation" of any sound is selected randomly from among several in the category assigned to the particular noise the dog is making. Thus a sound analyzed as "on-guard" might be translated as "You're not so tough!" one time, "I might bite!" the next, or "Enough is enough!" on another occasion.

Within the context of analyzing emotions, the device works moderately well. I programmed myself as a "large dog" with a "short snout" and then growl-barked in a low pitch. The Bow-Lingual interpreted this as an on-guard emotion meaning, "I don't like you either!" On the other hand, repeated high-pitched whimpers and whines were interpreted as frustration, first with the phrase, "Please be nice to me," and then a moment later as, "Something's bothering me." However, when I gave my best imitation of an alerting bark (as for something unusual nearby), it was interpreted as a happy emotion and translated first as "Aren't I great?" and a second time as "I'm gorgeous!" Perhaps dogs in Japan have a different accent that I can't mimic.

Although Bow-Lingual was designed as a means of getting to understand a family pet, a variation of the idea of electronic translation of dog language has been successfully designed as part of a security system. It was introduced by Eyal Zehavi, founder of Bio-Sense Technologies in Israel. Using dogs as guards is nothing new, since it is understood that dogs have better night vision than humans, plus significantly better senses of smell and hearing. However, when Zehavi was working on military bases, he found that that human security personnel did not always understand what a particular dog's bark meant. They often dismissed the dog's barks as unimportant or insignificant—perhaps simply a response to a cat or another dog passing nearby, and because of that break-ins still occurred.

Zehavi's researchers used computers to analyze 350 different dog barks, and his summary of their findings was that, "Dogs have a specific bark when someone threatens their space. It doesn't matter what breed of dog they are, how big or small, or what sex; all that matters is that they bark in response to a threatening situation. An alarm bark is always the same."

On the basis of this, Bio-Sense Technologies created the DBS, which stands for *Dog Bio-Security System*, which can be integrated into existing security systems. Like Bow-Lingual, it has a sensor mounted on the dog's collar that conveys information to a receiver, where a human guard may be monitoring a number of dogs. The data is filtered and passed through an analytical processor to determine the state of the dog. This procedure can include additional channels of information, such as the dog's heart rate, in order to make interpretation clearer and avoid false alarms. The resulting data are translated into three levels of alert, similar to the national security system, which uses colors.

Here, we might have a "Code green" or "no-alert" status, meaning that the watchdog is responding to routine events, a "Code orange" or "medium alert," meaning that there is suspicious activity in the vicinity, and a "Code red" or "high alert," meaning that an intrusion has occurred into the dog's territory.

Although DBS can cost many thousands of dollars, it is still only one-quarter the cost of video surveillance systems. As of this writing, it has been found sufficiently accurate and sensitive to do the security work required, and for this reason the system is now being used in Israel's high-security Eshel Prison, as well as at some Israeli military bases, water installations, farms, ranches, and garages, and in Jewish settlements in the occupied West Bank.

I spoke to a colleague of mine about the Bow-Lingual and some of its spin-offs, like the Dog Bio-Sensor Security System. He mentioned to me that a former student of his was working on an electronic translation system that was much more grand and elaborate. It was supposed to ultimately be the equivalent of the "universal translators" depicted in many science fiction shows like *Star Trek*, namely a computer-based device to translate communications from any language, even if you personally could not recognize which language the message was in. The theory behind the research was that in all human language systems, the sounds that indicate emotions have certain similarities, and the sounds associated with certain people, events, and things have notable commonalities. For example the word for "mother" in many languages has a prominent "m" sound, as in the French *Mere;* Italian *Madre;* Serbian *Majka;* Dutch *Moeder;* Estonian *Ema;* Russian *Mat;* Greek *Mana;* Hindi *Maji;* and even in fairly esoteric and isolated languages, such as the Hawaiian *Makuahine;* the Urdu *Ammeô;* or the Swahili *Mzazi.* The translation project was being financed by the United States government, and my colleague suspected that ultimately the goal was to use it for automatically monitoring and translating phone calls and perhaps audio Internet messages, as part of an anticrime or anti-terrorism program.

My colleague then mused: "Since all animals use much the same variations in sound quality in their communication, I wonder what would happen if a sample of dog sounds was put into the translator?"

I got in touch with his former student who was working in a government research facility in Bethesda, Maryland. He was surprised to hear from me since the project was considered to be "internal" and details were not being widely broadcast. The idea of trying to translate the language of dogs intrigued him. He added some cautionary notes, however.

"First, you must understand that the translator is a work in progress and is still far from perfect," he explained. "It does not interpret single words, per se, but rather tries to put them in the context of the words before and after them. It then comes up with a set of candidate words, and selects the one that it feels is most likely. This means that you often get ungrammatical sentences, but at the stage that we are now at, you can at least pick out the general meaning of what the person is saying. Because it requires a context, I need an utterance that goes on for at least thirty seconds to work on."

Thirty seconds is a long time for a continuous bit of dog language; however, there was one dog that I had lived with for a short time that "talked" all the time. She was a Cavalier King Charles spaniel named Amy, who became mine when she was well past seven years of age. She was to be a gift for my parents, who wanted an adult dog. Amy stayed with me for nearly a month while I arranged for her transportation from Vancouver to Philadelphia. She was the ultimate dog orator, and would walk around the house giving canine monologues—combinations of barks, whines, little yodeling sounds, low-key growls, and so forth. Just for fun I taped one of her "speeches" to play over the phone for my parents, with a comedic explanation that this was Amy's list of what she wanted them to do to prepare for her arrival.

I dug up the tape, which for some reason I had kept, and sent it off to be translated. A week or so later I got an email, which said, in part, "When I put it through the translator, at first it didn't make any sense to me. However, when I looked at it a while, it became clear that it wasn't just straight prose, but was really poetry. So I took the liberty to parse it out the way I think it goes."

This is Amy's poem:

Happy closets speak falling biscuits to lazy cats.
Shadow puppies dance sunshine music.
Warm dreaming goes loudly through days.

I'm not convinced that there is anything valid in this "translation"; however, a free-form poem about biscuits, cats, puppies, and daytime naps could well be what an old dog might sing about as she wandered happily and "loudly through days."

Chapter 17

Are Dogs and Cats Incom*pet*able?

ONE HOT SUMMER day in my youth, my grandfather Jake attempted to explain one of life's great mysteries.

"Dogs and cats started out as the same animal—only the dogs were all men and the cats were all women," he said. "Now that's the real reason dogs chase cats!"

Even my six-year-old mind found that story hard to believe, but many widely-held beliefs about the relationship between dogs and cats are not much more credible than my grandfather's folk tale.

Most people firmly believe that dogs and cats are traditional enemies and coexist only with great difficulty. If a couple is described as "getting along like cats and dogs," we know that means they are fighting or nasty toward one another. Yet, according to the most recent U.S. survey data, 36 percent of American homes have at least one dog, and 54 percent of these homes also have a cat. This would not be the case if dogs and cats were not getting along reasonably well with one another.

People have come to believe that dogs naturally dislike cats because we often see dogs chasing or barking at cats. There are two explanations for this chasing behavior. The first is evolutionary. Cats' normal response to a potential threat is to dash away and find safety. Dogs, on the other hand, are cursorial hunters, which means they are built for swift running. When they see something moving quickly, their instinct is to chase it, whether it is a squirrel, rabbit, bird, ball, Frisbee, or cat. This means that the cat's instinct to run is an automatic trigger for the dog's instinct to chase.

The second, more subtle reason is that cats and dogs tend to misinterpret each other's body language. Most of the facial signals that dogs and cats use are quite similar. For example, they open their mouths and show their teeth when they want to be threatening. They move their ears up and forward to indicate interest, and slick them down or to the side when they are fearful. Aggression is one "ear" message that is subtly different in cats and dogs. Normally a dominant aggressive dog will hold its ears erect and forward. Just before initiating an attack, the ears might be tilted slightly out to the side, widening the apparent "V" shape that the upright ears make. For the cat, this gesture is much more pronounced, and the ears turn, so that the inside surfaces are directed down toward the side and the backs of the ears are now visible. In some of the larger wild members of the cat family, the backs of their dark ears are actually marked with distinctive patterns, which makes this rotation of their ears more conspicuous when they are signaling their aggressive intentions. Other wild cats have tufts at the ends of their ears, which also help to make the ear's positions and rotations more visible from a distance. This is a useful signal to be aware of, and I have

Ears held flat like airplane wings are an aggressive threat in cats.

taught my grandchildren the simple rule that, "When a cat's ears turn into airplane wings, you should stay away because he is grumpy or angry."

However, dogs' and cats' eyes give different signals. Cats normally open their eyes more widely than dogs and tend to stare at anything that approaches them. For dogs (and humans, for that matter), a wide-eyed stare is considered a threat, often an expression of dominance, and it is clearly meant to intimidate the individual being stared at. Thus consider the situation where a dog is approaching a cat that is merely being attentive and not trying to be hostile. The dog can easily misinterpret a cat's wide-eyed gaze as a challenge. When the dog responds by adopting a threatening posture or barking, the cat runs away and then of course the chase is on.

Dogs and cats also often misinterpret each other's tail signals. For a dog, a broadly wagging tail, moving from side to side, is a nonthreatening, submissive gesture that signals friendliness and approachability. For a cat, however, that same tail movement is

An attentive cat's wide-eyed stare can be misinterpreted by a dog as a threat.

part of a threat or hunting signal. This is why a dog that sees a cat wagging its tail will approach to greet it, only to get a face full of claws. The dog now thinks, "That cat lied to me, so I will never trust a cat again."

A similar misunderstanding can come about when tails are held upright and high. For dogs, the tail held straight up, or curved somewhat over its back, is a sign of confident dominance, and an assertion of authority that will be backed up by a show of aggression if necessary. For cats, however, this same signal is one of the friendliest signs that the species ever offers. The high tail, with perhaps a slight forward curve over the back, allows another friendly cat to investigate the exposed region under the tail. Cats, just like dogs, have scent glands around their anal region that provide pheromones that can identify the individual to anyone who catches the familiar odor. Raising the tail is much like offering someone your passport or driver's license to verify your identity. A dog that sees this upturned tail signal from a cat is apt to misinterpret this cattish message of friendship as an attempt to express dominance. The cat may similarly feel betrayed when its offer of friendship is met with suspicion and threat and may never trust a dog again.

There is another, different, meaning for a vertical tail position in cats. In this case the tail held high is only one of several signals being sent at the same time. A frightened cat bristles its fur, arches its back, and holds its fluffed out tail erect and vertical. This posture, which produces the typical Halloween cat silhouette, means that the cat is very frightened about the current situation. In dogs this same *piloerection* (the technical term for the bristling or fluffing-up of an animal's fur) is designed to make them look larger and represents a high level of aggression. As the dog's tail bristles out, he raises it into a dominant and threatening vertical position. When a cat sees the fluffed-out fur and high bristling tail of a dog, he may misread this as equivalent to the cat body language for "I'm terrified." The cat may thus actually trigger an attack by failing to back down to the dog's threatening signal. The dog can just as easily misinterpret the cat's indication of fear. If he reads the cat's signals as if they were dog-body language, he would interpret the bristling fur and a raised tail as meaning that the cat is challenging him and is insisting upon a fight. It probably only takes one misinterpretation and

A friendly, inviting cat meets a dominant, assertive dog.

an unexpected bite or scratch to create a lifetime of hostility between the two species.

The good news is that dogs and cats living together can learn to interpret each other's signals. A dog will come to think that its own feline housemate is a bit odd, with poor language comprehension and communication skills, but definitely not evil. Other cats that he does not know, however, will remain liars and enemies to be chased and chastised.

Because dogs are basically social animals, they can come to accept a cat as part of their family. Learning to live together works best if a cat is introduced into the home where the dog lives rather than the other way around. The reason for this is that cats are often quite territorial and will view the dog as an interloper that will try to take over his kingdom. Dogs, on the other hand, view new arrivals to their home as potential new pack members. A dog that has never been hurt by a cat will usually attempt to greet the new arrival in a friendly manner, rather than attacking it. A cat's acceptance of a new housemate may take much longer than a dog's.

When a new pet is a young puppy or kitten, both species tend to respond more positively. In fact if a puppy or kitten is brought into a home where an adult of the other species lives, it can even learn some of the other's behaviors.

The Collins family in Philadelphia learned this when they rescued a mixed-breed terrier puppy, named Flash, that was only four weeks old. His owners introduced him to their cat Mildred, the mother of a new litter of kittens. They hoped the kittens would provide Flash with some company. Flash was the same size as the kittens, and Mildred adopted him as if he were one of her own. She even cleaned him with her tongue, in the same manner that she groomed her kittens.

By the time he was sixteen weeks old, Flash's behavior was more like that of a cat than a dog. His favorite toys were cat toys, like a squeaky mouse and a ball with a bell in it. He even learned cat mannerisms, such as the typical postures in stalking and pouncing when playing with his adopted feline family. Perhaps most surprisingly, Flash also learned the typically feline habit of washing his paws with his tongue and then using them to clean his face and ears. And when given a choice, he always selected the company of cats over that of other puppies.

Sometimes, the relationship between a dog and a cat forms instantaneously, and a lifelong friendship results. For instance, one cold day in Bristol, England, a gang of boys stole a kitten, threw it into a pond, and waited to watch it drown. Suddenly, a Labrador retriever named Puma dashed into the water to grab the kitten.

Puma must have thought this was some sort of accident because he brought the kitten out of the water and laid it at the boys' feet. They just laughed, kicked him away, and threw the kitten back into the water.

Puma again leapt into the water, but this he time swam across to the other side of the pond with the kitten held softly in his mouth. Once ashore, he ran home with it.

When Puma's family opened the door, he rushed in and laid the kitten next to the heat vent. He simply would not let his family take the kitten out of his sight, so they decided they had to keep it. They named the kitten Lucky.

Lucky and Puma spent many years curled up together and napping in the sunshine. When I think of canine-feline relationships like this, I remember my grandfather's folk tale and wonder—if cats and dogs can learn to live together peaceably, why can't men and women?

Chapter 18

What Dogs Can Teach Kids

THE HUMORIST ROBERT BENCHLEY once observed that "A dog teaches a boy fidelity, perseverance, and to turn around three times before lying down." It is only recently, however, that science has caught up with Benchley and demonstrated that there are many things that children can learn from living with a pet, particularly a dog.

Young children seem to be naturally attracted to pets. Companion animals quickly catch the attention of kids. They like to watch them and are curious about what the animals are doing and whether they can communicate with them. This has been demonstrated many times, such as in the classic research by psychologist Aline Kidd of Mills College in California and her husband, Robert Kidd, who studied children and their pets. They observed how infants and toddlers, ranging from six to thirty months of age, behaved toward their pet dogs and cats compared to their behavior around a lifelike, battery-operated toy dog or cat. The babies smiled, touched, followed, and made sounds that appeared to be attempts to communicate with the live animals (especially the dogs) much more than they did to the toy ones. Psychologists suggest that the attractiveness of animals comes from the fact that their behaviors can be both predictable and unpredictable. Much research also shows that children's learning is best when it occurs within a meaningful relationship, so because of their immediate positive feeling toward animals, kids are motivated to find out what their pet will do next—to make the behaviors more predictable.

At some level, parents recognize that the family dog has a special status in the family and a special relationship with their children. For example, one study conducted in Germany supported the idea that parents believe that their kids are learning something from their pets. In fact 90 percent of parents surveyed thought that the family dog played an important role in teaching their young children social skills and improved the children's quality of life. This study confirmed the special status that pets have, since it found that 80 percent of the children interviewed considered their dog to be an important friend and confidant. In analyzing their research, psychologists June McNicholas and Glyn Collis of the University of Warwick in England found that, when it comes to children's social networks, "Cats and dogs frequently ranked higher than many human relationships."

McNicholas did another research survey that demonstrated that a pet dog quickly becomes part of a child's social and psychological support system. Using a test group of 338 children, she found that 40 percent of the children sought out their pets' company when they were upset and needed reassurance. Social interactions with the family dog were also very important, and 40 percent of the children looked for interactions with their pet when they were bored; and 53 of the kids liked to watch TV or videos with their pet. Overall, 85 percent of the children tested regarded their pets as a playmate. Just how important pets are in the lives of children was confirmed in other research. When a group of seven- to ten-year-olds were asked to name "the ten most important individuals in their lives," their family pets virtually always made the list. Dogs were often listed as more important than the child's brothers and sisters.

This caring relationship for a pet translates into learning in the same hit-or-miss fashion that characterizes most social learning. Children learn that if they pinch or pull the tail of their pet, it becomes distressed and runs away. If they pet it, speak nicely and quietly, and give it treats, their dog comes to them, plays, and acts happy. Seeing how others respond to our behaviors is the process by which we develop "empathy," the ability to recognize, perceive, and indirectly, even feel the emotion of someone else. A person with highly developed empathy can "put himself into another individual's shoes," meaning that he can more effectively interpret the moods, thoughts, and emotions of others.

Many studies confirm that children with pets have higher degrees of empathy. In research done at the University of New Mexico, Robert Bierer studied kids aged ten to twelve years and considered the effects of dog ownership on the children's social skills. He summarized his research this way: "People have known for years that dogs are good medicine for children. What I found is that preadolescent children with pet dogs have significantly higher self-esteem and empathy than children without dogs. These higher ratings in self-esteem and empathy hold true whether the dog is 'owned' individually by the child or by the entire family. That means that just having a dog in the house makes a difference, regardless of whether the family is headed by a single parent, the mother works outside the home, or the child has siblings."

A study of 455 schoolchildren between the ages of eleven and sixteen extended this research by isolating one of the skills behind this increased empathy. In order to interact with a pet, you need to understand its emotional state, its desires, and its intentions. Obviously, since pets don't talk, this requires careful observation of the animal's behaviors. Many signals in the body language and facial expressions of dogs are somewhat similar to those in humans. Certainly the experience gained by observing social signals from a pet dog may carry over to interactions with humans, to allow faster learning of these subtle signs. This was verified in research that found that children with pets (dogs and cats) were better able to understand nonverbal communications in general. These children not only understood dogs better than nonpet owners, but also more accurately interpreted the body language and emotional expressions of the people in their lives.

The increased self-esteem and feelings of competence reported

Children with pet dogs learn to read body language better, and this skill carries over to their interactions with the people in their lives.

by Bierer and a number of other researchers comes from the very nature of pets, which depend on human care for their survival and happiness. Pets provide children the opportunity to learn about, practice, and become interested in, nurturing living things. The need to nurture, help, or care for others seems to be a basic human instinct, even in young children. Having a pet serves as an outlet for this urge in young kids, who might not have any other opportunity to do so. Research shows that kids without younger siblings spend more time caring for and playing with their pet, and 92 percent of eight- to ten-year-olds who shared the responsibility for pet care felt that this was an important and special part of their relationship with the animal.

Studies conducted at Oregon State University demonstrated that nurturing and caring for pets has real psychological benefits for children. As part of the study, researchers taught a group of preschool kids how to look after a puppy. When tested later, the children regularly involved with caring for the puppy ended up being more socially competent. Generally speaking, these kids were more effective social communicators since they were better able to understand other children's feelings and intentions. In addition these children felt better about themselves and more in control of their lives. An interesting, unexpected benefit was that these children also became more popular. Dr. Sue Doescher, a psychologist involved in the study, explained the results this way: "It made the children more cooperative and sharing. Having a pet improves children's role-taking skills because they have to put themselves in the pet's position and try to feel how the pet feels. And that transfers to how other kids feel."

Elizabeth Ormerod, chair of the Society for Companion Animal Studies, summed up our knowledge to date saying, "For many years, the valuable role of pets in children's development has been recognized. But recently, the positive health, educational, and therapeutic benefits of pets have been scientifically investigated and acknowledged. Children tend to form very special attachments to companion animals. Through learning to care for and understand animals, children develop greater empathy for other people, which helps to improve human and animal welfare in society."

There is a potential downside to bringing a pet into your home when you have children, however. Not only is the interaction between the child and the pet important, but you as a parent are also going to set an example. Companion animals are not inanimate objects but living, feeling beings that you choose to bring into your life. Suppose that you give your child a pet on a whim, but then later decide it is too much trouble, and cause the animal to "disappear" by sending it to a shelter or rescue group. In effect, you are teaching your children that emotional relationships are cheap and expendable. If you continue to live with the pet, but show your frustrations and impatience through heavy-handed training methods and punishment, you teach your children that violence is an acceptable means to an end. Research confirms that children who have lived in families that got rid of their pets treat emotional relationships as if they were less permanent and less valuable. They model themselves after the actions of their parents, and when they become adults they are less likely to form an emotional bond with their own pets and more likely to give their pets up to a shelter. Similarly, children who grow up in families where pets were routinely punished and harshly treated are more likely to use similar forceful methods on their own pets and their own children later in their lives.

If you bring up your dog by showing care and understanding for your pet, however establishing clear boundaries for behavior yet still having room for unconditional love, you are drawing an emotional blueprint for your children that they can follow all the days of their lives. This raises the question then: "Is the dog teaching the child, or is the parent teaching the child with the help of the dog?"

Chapter 19

Confidants to Kings and Presidents

WRITTEN HISTORIES HAVE a way of focusing only on the leaders, such as kings, emperors, generals, presidents, politicians, and wealthy, powerful business moguls. They tend to ignore most of the advisors, confessors, confidants, friends, comrades, and soul mates of those leaders unless they are involved in some significant betrayal. These unsung individuals often comforted those history makers in their times of trouble, shared their secrets and their insecurities, reveled with them when things went well, and provided love and friendship to sustain them when the going got rough. Such companions are even less likely to be noted by historians if they had four legs and a tail.

Throughout history, leaders have felt isolated. Those in power often worry about exposing their concerns to others for fear that their secrets might be deliberately used against them, or inadvertently become known to enemies who might take advantage of them. For this reason many people in power have turned to dogs as their most intimate companions. Take the case of General Dwight D. Eisenhower, who would go on to become president of the United States after World War II. In 1943 he was Supreme Commander of the Allies' military operation and was coordinating the operations in North Africa that would finally wrest control from the Germans. He wrote to his wife, Mamie, "The friendship of a dog is precious. It becomes even more so when one is so far removed from home as we are in Africa. I have a Scottie. In him I find consolation and diversion . . . he is the 'one person' to whom I can talk without the conversation coming back to the war."

The Scottie that Eisenhower mentioned was Caacie, the dog that accompanied him back to England and continued providing companionship during the general's planning for the D-Day invasion.

Many American presidents have had special relationships with dogs. One of the best known was Franklin Delano Roosevelt, whose Scottish terrier, Fala, went virtually everywhere with the president. He regularly slept with FDR and attended most cabinet meetings. Fala was aboard the U.S.S. *Augusta* when Roosevelt and Winston Churchill (with his own constant companion, a poodle named Rufus) met to sign the Atlantic Charter. The only time that Fala found himself rejected was when he attempted to jump into the car beside his master, who was preparing to leave for his third inauguration. Since that seat had been reserved for the Speaker of the House, Roosevelt laughed and instructed the head of the Secret Service contingent, "Would you kindly check the credentials of this individual? If he does not have an invitation to be on the Inaugural platform, please have him removed."

A few years later, however, as Roosevelt had wished, Fala was buried at Hyde Park, New York, to finally rest again beside his beloved master, as he had done all of his life.

Lyndon Baines Johnson loved the companionship of his dogs so much that he had Christmas cards made up with a picture of his beagle, Him, and his white collie, Blanco, standing next to the president. Each card was signed with Johnson's signature and the paw prints of the two dogs. When his daughter, Luci, was married at the White House, Johnson planned to have the dogs be part of the ceremony. His wife, Lady Bird, refused to allow this; however, the president still managed to sneak the dogs into the proceedings as the family posed for the official wedding pictures.

John F. Kennedy's presidency included nine dogs, but his favorite was a Welsh terrier named Charlie, a scamp that would silently slip up behind any gardener digging or working the soil, then make a quick rush at the man, nipping him in the seat of the pants or grabbing at his leg. Once, when a worker complained, he was told to forget the incident since JFK was more likely to have the man dismissed than to take action against the dog. Charlie was a source of comfort to the president, as was proved during the Cuban Missile Crisis when JFK interrupted the deliberations of his advisors to have Charlie brought to him. For what seemed like a very long time, Kennedy sat stroking the dog, and the president gradually seemed to relax. Then, with a calm look of control he put Charlie down and said, "I suppose that it's time to make some decisions."

Many other presidents have had significant relationships with their dogs, including Ulysses S. Grant, who appointed his Newfoundland dog to the post of White House steward, and George Bush Sr., who told me that during his presidency he would often be joined during his morning shower by his springer spaniel, Millie.

It often appears that dogs actually perform useful tasks for people in leadership positions. In the case of Kennedy and the missile crisis, his dog Charlie obviously reduced the president's stress and anxiety, in much the way that a cherished confidant, advisor, or special friend might do. History records that dogs have served this particular function many times, bringing solace to significant people during times of pain or crisis and often serving the same psychological functions that a family priest might fulfill. For example, Mary Queen of Scots spent her long hours of prison confinement in the Tower of London in the company of only her small spaniels. She spoke to them, and they comforted her through the night. In 1587, when she was beheaded, it was found that she had hidden one of the small dogs under her voluminous robes. Afterward, according to one eyewitness, it "would not depart from the dead corpse" and had to be carried away. Many years later the person who ordered Mary's execution, Queen Elizabeth I, who also had few intimates to ease her loneliness, spent her last night "counsolled only by her dogge," a very similar toy spaniel.

One of Elizabeth's direct successors eventually gave his name to the breed of spaniel that she and her half-sister Mary both loved. Like his predecessors, Charles II of England also slept with his Cavalier King Charles spaniels and even had them included in a ceiling mural in his bedroom in the palace at Holyroodhouse. Charles usually had one or two of his dogs with him at all times, which led to an odd incident. As the king was about to enter the House of Lords one day, the Sergeant at Arms informed him that the dogs could not accompany him. "Only Lords may enter, my Liege," he stated. The king promptly issued a decree conveying an hereditary title upon his dogs, making all of that breed members of the peerage. Not only does that entitle them access to the House of Lords, but in theory, since the decree has never been revoked, should a Cavalier King Charles spaniel scratch at the gates of Buckingham Palace today, by established tradition, it must be granted entry.

Although the dogs of powerful leaders generally have had the sense to steer clear of political involvements, at least one dog served as an intermediary or ambassador and ultimately saved a man's life. Czar Peter the Great of Russia had few close relationships, but he had a strong affection for his Italian greyhound, Lissette, who shared his bed more often than his wife, Catherine. It happened that one of the members of Peter's court was accused of corruption. The accusation would later prove to be false and malicious, but when Catherine was apprised of the circumstances and attempted to intervene on the accused man's behalf, Czar Peter (not known for his calm demeanor) flew into a violent rage, feeling that it was inappropriate for Catherine to involve herself in political matters. He forbade her ever to mention the case again in his presence. Distraught at the mounting evidence of the man's innocence, Catherine wrote a message to Peter, petitioning for clemency, signed it with Lissette's name, affixed the dog's paw print, then tied this note to Lissette's silver collar. Later that evening, when Peter was preparing for sleep, he found the message. He sat on the edge of his bed gently petting Lissette's head and then, without further comment, called for his secretary and drafted a pardon that night. The matter was never mentioned again by Catherine or Peter.

Stories involving interactions with dogs could be told about hundreds of famous and powerful people. The list includes: Alexander the Great, Robert the Bruce, George Washington, and the entire Ming dynasty of China and its emperors, to name but a few.

Up to now we have only seen the idyllic side of dogs acting as advisors and confidants and never betraying their powerful owners' secrets. Yet such is not always the case. Take the case of Napoleon Bonaparte's wife, Josephine. Following their whirlwind courtship and swift marriage, Napoleon was chagrined when he learned that Josephine already had a bed companion whom she was unwilling to evict—her pug, Fortune.

Josephine and her pug, Fortune.

Fortune was a particular favorite of Josephine because he had proven helpful at a time of crisis during the revolution. When the revolutionary council had imprisoned her first husband, Alexander, because he was an aristocrat, Josephine was also imprisoned and under threat of execution. Her children and others were barred from communicating with her; however, no one paid any attention to the little dog that was brought to visit her each day. By hiding messages under Fortune's broad velvet collar, Josephine maintained communication with those outside the prison. Several of the outgoing messages were for highly-placed officials whom Josephine knew, and who managed to intervene to delay her execution. When Robespierre, the most radical of the revolutionaries, fell from power, these friends obtained her release.

On her wedding night with Napoleon, Josephine insisted that the dog remain in the room and apparently on the bed. Later that evening, as Napoleon made love to his new wife, they were interrupted when the dog attacked, sinking his teeth into the calf of the naked and otherwise-involved general. The wound was sufficiently large and deep that he bore its scars for the rest of his life. It also left him harboring an intense dislike of Fortune and all pugs.

Soon after, Napoleon had to leave his wife alone in France while he took command of the Army of Italy. Shortly after he left, Josephine started an affair with a young army lieutenant named Hippolyte Charles. When rumors began circulating, Napoleon felt that at least for the sake of appearances, he would have to bring his wife to Milan. Under pressure, Josephine agreed to leave France, and a six-carriage convoy was put together to move her household and family guards to Italy. Josephine sat in the first coach, accompanied by Fortune, Lt. Charles, and Napoleon's brother, Joseph. Joseph and others on that journey would later tell Napoleon that Fortune behaved unusually well during this trip. This was remarkable because the dog had a reputation for being nasty and snappish toward almost everybody except Josephine. Yet he showed no negative behavior toward Lt. Charles, and the obvious conclusion was that this young man had been with Josephine intimately and over an extended period of time. Josephine did not usually take Fortune with her to formal or even casual social functions, but the dog was always in her bedroom and on her bed when she slept. Clearly, Lt. Charles may well have achieved his friendship with the dog while in Josephine's bed as well. In modern terms, psychologists would say the pug had become habituated to a man he would normally see as an intruder in his territory. Based on the dog's behavior, Napoleon started a series of inquiries that eventually proved that Josephine had been cheating on him, and their marriage began to unravel.

Josephine had trusted her dog, and he had unwittingly betrayed her confidence. Even when relying on the companionship of dogs, it appears, the famous and powerful must exercise discretion.

Dogs and Modern Society

Chapter 20

Medicine for the Mind

IN THE AFTERMATH of Hurricane Katrina, with virtually everyone who wanted to leave the city of New Orleans either rescued or evacuated, television screens carried pictures of other victims of the disaster—dogs stranded on rooftops looking down at the floodwaters. One video clip showed a dog swimming through the foul water, desperately trying to reach a rescue boat after its owners were forced to abandon him. Other scenes showed sad, starving animals on balconies or staring despairingly out of windows. Such mournful sights stirred the emotions of many who saw them, and questions began to be asked.

At one press conference, Michael Brown, the director of the Federal Emergency Management Agency (FEMA), who ultimately proved to be lacking in the competence to adequately deal with a calamity of this scale, was asked by a reporter, "What about the dogs and cats that have been stranded?" His response began, "They are not our concern . . ."

A short while before Katrina hit, FEMA had gone through a disaster preparedness exercise which involved a mythical hurricane, *Pam*, hitting the U.S. Gulf Coast. Extensive computer simulations and hands-on practice by search and rescue, police, military and civil authorities, engineers, and medical experts were involved. When Ivor van Heerden, a hurricane researcher from Louisiana State University, who helped direct the simulation exercise, was asked about preparations to save pets, he answered, "They were not part of our plans because they are not considered to be important."

The events surrounding the real-life disaster that followed proved such planners to be wrong. It demonstrated that many people who live with animals consider them important enough to risk their own personal safety to keep their pets from harm. The rescue planners had forgotten that saving the human body is not enough. People need affection, comfort, family (or something that serves as family), as well as a feeling of being needed. These emotional needs often must be met before people can motivate themselves to try to survive physically. For many people, such requirements are filled by the companionship of an animal. Pets are part of their families, and such people would no sooner think of abandoning them than abandoning a child. One exhausted National Guard officer explained to General Russel Honore, who was coordinating the rescue efforts, "We estimate that 30 to 40 percent of the people who refuse to leave the affected areas are staying because they want to take care of their pets."

In the early days of the rescue effort, some authorities showed an incredible degree of callousness. Since no planning had been done to take care of pets, people were simply ordered to abandon them. A heartrending example involved one young boy, among the thousands who ended up being sheltered at the Superdome. When he tried to board a bus to Houston while carrying a small white dog, a police officer snatched the dog from the boy. This little animal would not have taken away any space needed by a human survivor. As it was carried away, the boy sobbed "Snowball! Snowball!" then, overcome with his distress, he sank to his knees and vomited. One woman, with no other possessions left, offered her rescuer the wedding ring off her finger to save her dog, but to no avail. There were even stories of local authorities in St. Bernard Parish, who, rather than argue with survivors about saving their dogs, simply shot their pets.

Some rescuers, however, found room in their hearts for compassion and some means of helping. Many of the National Guard personnel left water and food for stranded dogs in the hopes that they would survive long enough to be saved. Louisiana State Treasurer John Neely Kennedy was helping people board buses near Baton Rouge and found himself intervening when some evacuees resisted because they had been ordered to leave their pets behind. One woman pleaded, "I've lost my house, my job, my car, and I am not turning my dog loose to starve." Kennedy joined other volunteers in taking down the names of those leaving on the buses and asked the Louisiana SPCA to come out and collect the animals. Soon it became standard practice for representatives of the Humane Society of the United States and the ASPCA to meet people brought in from the flood, taking their animals to shelter while recording information so that people could later be united with their pets.

Shortly after leaving the bus-loading area, Kennedy found a mixed-breed dog tethered to a post near the road with an unopened can of dog food next to him. With the dog was a plaintive note that read, "Please take care of my dog, his name is Chucky." Kennedy said "What else could I do? I am taking care of Chucky."

There are many stories of disaster victims who resorted to

The dog was tethered near the road with a note and an unopened can of dog food.

extreme measures to save their dogs. Take the case of Dohnn Moret Williams (who likes to be called Moret). His former home was completely submerged under the floodwater, his possessions gone, and his elderly father, who also lived in the city, was presumed dead. Still, sitting outside of the rescue center in the Houston Astrodome, which was his temporary shelter, he felt some relief. "I spent most of the morning crying when I knew I could come get him," he said as he reached down to pat Sebastian, a large black cocker spaniel with red markings above his brown eyes. Sebastian had just been reclaimed from the Houston SPCA. "I got no children. This here's my baby."

The exit of Moret and Sebastian from New Orleans was treacherous and exhausting. Surrounded by the polluted, sometimes neck-deep floodwater, Moret knew that the dog would not be able to swim the whole way to safety. So he found an air mattress, and although Sebastian didn't like it, because it tipped when he moved, it gave Moret something that he could tow.

Ultimately, they made their way to an elevated portion of Interstate 10 from which people were being evacuated by helicopter. Unfortunately, the rescuers were under orders to prevent pets from boarding.

"There was no way I was leaving without him, and I figured I would do whatever I had to in order to keep him with me," Moret said. "I got a big black trash bag and put Sebastian in it. Then I whispered to him not to make noise."

Surprisingly, the dog seemed to understand. There was, however, one moment when the whole scheme looked as if it were going to fall apart. Squeezed up near the front of the helicopter on Moret's lap, the dog started squirming. Moret said, "He bumped against the pilot, and I thought it was over, but the pilot just goes, 'I didn't see nothing.' " Their subterfuge was not yet done. Moret was given a ride to Houston on a bus that also was under orders not to accept animals. This time, safely seated toward the back of the bus, Sebastian made the whole trip with his nose sticking out of the top of the bag. When the two arrived at the Astrodome, volunteers from the SPCA were waiting. Sebastian was only one of many stowaways on the buses. Some dogs were carried in bags or suitcases, and some even hidden under flouncy blouses or stuffed down baggy pants. All were given temporary shelter until their owners reclaimed them. Moret and "his baby" were together again and went off to stay with his sister.

Sebastian and Moret were lucky. Many other dogs would not make it through this tragedy. Numerous pets were left behind during the initial evacuation of the city, many with food and water, since caring owners hoped they would be away for only a few days.

These events point out an important set of guidelines for people who live with pets but are facing an emergency situation. First, at a minimum, each animal should wear identification, such as a small metal cylinder that hooks on to a collar and holds a slip of paper. On the paper you should record the dog's name, your name, address, telephone number, and e-mail address. A cell phone number or an out-of-town contact is also helpful, in case your city or neighborhood is devastated to the degree that local contacts are unreliable.

Second, if at all possible, people who share their homes with dogs should never leave them behind in an evacuation. The truth is that you don't know when you'll be able to return to your home and when, or even if, humane agencies will be allowed to rescue your pets—presuming that they survive the initial emergency. Simply put, if you have the means to evacuate, your dogs are safest with you, even if it means you have to camp out. Traveling with your dog in times of crisis may slow your progress, and you may have to make compromises to keep your pet with you.

Fortunately, there are usually people among the rescuers who understand that dogs are not just property to be abandoned like extra baggage. They understand that dogs fulfill an important psychological function and may be a survivor's only link to affection and the life they used to live.

At one stage in the rescue efforts, an elderly woman was getting ready to board a helicopter to be evacuated out of New Orleans. Against her breast, she hugged a little Yorkshire terrier. At the door, an enlisted man took the dog and said, "Sorry, Ma'am, but orders are 'No animals.' "

The woman's weary eyes filled with tears, "I got nothing and no one. He's all I got left!" The soldier stood there holding the dog and repeated, "Orders are 'No animals.' "

At that moment, an officer wearing captain's bars and medical corps insignia appeared at the door. His name tag read "Anderson." He reached over to the enlisted man and took the little animal. "That's not a dog," he said, "that's medicine."

"Medicine?" asked the puzzled soldier.

"Medicine for the mind," said Captain Anderson, as he

handed the dog back to the woman and helped her through the door.

There is a hopeful postscript to this story. I was one of several psychologists, dog experts, and humane society officials who never wanted to see this sort of tragedy occur again. We prepared a very short, simple, commonsense set of guidelines such as those that I mentioned above, concerning dogs and other pets in times of natural disaster. These were to be given to emergency workers, National Guard troops, and people forced to evacuate because of some form of crisis. We had just distributed the first wave of these when Hurricane Rita struck several U.S. states bordering the Gulf of Mexico. Officials and the public listened to what we said, and people took their dogs with them when they evacuated. Although there were floods and devastation from this hurricane as well, at least there were no heartrending pictures of abandoned dogs following that storm. Now let us hope that people remember for the next time.

Chapter 21

The Lion Dogs of Buddha

IT IS SAID that in the year 69 A.D. the Emperor of China, Ming Ti of Han, returned to his palace from a visit to the White Horse Temple in his capital city, Luoyang. When he entered his throne room, he was preceded by two Pekingese dogs that barked as if to clear the path, and some claimed that two similar dogs followed him, carrying the train of his robes in their mouth.

When Emperor Ming reached the raised platform where his throne stood, he turned to face the gathered audience. He began by pointing to the four dogs that now spread out around him and he then announced, "These are my attendants. They are the *rui shi* ["auspicious lions"] who are the defenders of the imperial dharma and symbols of the faith."

The path that led the Pekingese dog to become a symbol of

royal status and privilege, as well as a symbol of Buddhism in China, is convoluted. It all began with a dream that Emperor Ming Ti had in which there appeared the striking image of a golden man. The next day Ming Ti asked his ministers to interpret this dream, and one, Zhong Hu, explained that the emperor's dream most likely had been about Buddha, a holy man in India. The emperor was impressed that someone could reach out over such a long distance to send him messages in his dreams. Exploring further, he formed a delegation of eighteen ministers and scholars to seek the secrets of Buddhism. After visiting Afghanistan, they eventually returned, bringing with them two eminent Buddhist monks who led a white horse that carried the holy writings or *sutras* of the religion and images of Gautama Buddha.

The emperor, who ultimately would be responsible for introducing Buddhism to China, learned as much as he could about the religion. The monks instructed him in many tales of Buddha, his exploits and teachings. During their instruction, they noted a deep association between Buddha and lions, explaining that the lion was a creature that Buddha often called on as a protector of the holy spirit, and whose image he sometimes accepted as his own. The lion is not indigenous to China and no one in the court knew what this animal actually looked like, but one of the monks wore a pendant that supposedly represented a lion. Unfortunately, the pendant was a flat medallion, so the face was squeezed down and represented by low-relief lines on a smooth surface. This distorted the shape of the lion's face into a widened mouth, short flat muzzle, and large eyes. For some reason most representations of lions at that time also stylistically elongated the ears so they hung down the side of the animal's face to some degree. As a result, the lion's face on this medallion resembled the face of the Pekingese dogs Ming's predecessors had received as gifts from Tibetan delegations.

Ming pointed out the similarity between the face on the monk's pendant and the face of one of his pet dogs. At the same time he could not believe that the lion that had been described to him as so very large, strong, powerful, and courageous could have any relationship to his pampered pets.

The monk smiled and told him that, indeed, there might be a closer association than the emperor imagined. To prove this point, he proceeded to tell Ming Ti this story:

"The great lion had been wandering the woods and accidentally stepped upon a long sliver of bamboo that drove itself into his paw. When he tried to pull it out, it snapped off, leaving most of it in his paw, where it caused great pain. His groans were heard by a little marmot, a sort of a ground squirrel, and she approached him saying, 'Noble companion of holy Buddha, stay still a moment and I will help you.'

"The brave little animal then approached, and with her sharp teeth was just able to grasp the remaining visible stub of bamboo in the lion's paw. She carefully removed the entire piece and gently cleaned his wound. The lion was not only grateful to the marmot but was quite impressed by her courage and spirit. He lay there for many hours talking with her and found that she was very wise in the ways of the world, and kind and spiritual as well. They met again several times, and soon it was clear that they had fallen in love.

"Such an unfortunate love! Although the marmot is big for a squirrel, it is only a fraction of the size of a lion, so their love could never be consummated. In desperation the lion approached Buddha and told him of his problem. Buddha thought about this carefully and asked the lion why their size difference should matter. The lion explained that his love for this marmot was so great that it should be sealed in a manner that would allow it to continue forever. In other words, they wished to have children together, to carry on a blending of both of their spirits for future generations. Perhaps, suggested the lion, if you could shrink me to the size of my lovely marmot, I could be happy.

"Buddha thought about this for a moment and replied, 'I still have great need for your strength and courage and cannot spare you for long. However, for a short time, something can be done. I will reduce your size for a period of only one circuit of the moon. During that time you will be small, but your abilities, courage, and spirit will remain unchanged. At the end of that month, however, you must return to my service in your normal size. If your union is blessed, then you shall have children. But after that month, your relationship with your lady love must be one of mind and spirit, not one of flesh.'

"It was as the lion hoped. He and his love spent a rapturous month together, and sometime after that she gave birth to a family of eight. Buddha was most pleased with these new animals, and when they were old enough he sent four of them to Tibet and another four to a far-off land in the East. There they were to serve as an example of the triumph of love and faith over all obstacles. Meanwhile, the lion and the marmot stayed together, sharing their union of spirit in their natural sizes for as long as they lived."

The monk paused and looked at the little dogs thoughtfully and suggested that it might be possible that these dogs, which after all were a gift from Tibet, might truly be the descendants of that original union of the lion and the marmot. He then suggested to the emperor, "If this is so, there is still the heart and the courage of the lion in these animals, despite their size. Because they are here by the direct intervention of Buddha, they are holy, and like their father should have the ability to ward off evil. Perhaps because they are small, their place in the world is to protect your palace and the temple from little demons, which because they are not large go unnoticed except for the damage that they do to body and soul. Because their heart is as great as a lion's and their will is strong, a dog like this could carry enough of the holiness from his heavenly conception to destroy such demons, even if the demons are fierce and perhaps even if they are large."

Pleased, Emperor Ming burst into laughter. "So, if I cannot have lions at my service then I will have Lion Dogs. They shall be part of my Imperial Guard, and wear the colors of my guard. They will have duties only to protect the holy places and the homes and persons of those of noble blood. No others can lay claim upon creatures with such an exalted lineage."

Small dogs had been popular among the nobility for hundreds of years before they became identified with the Lion of Buddha. They were a valuable commodity that was frequently imported from distant places and served as expensive and much-appreciated gifts. Around 500 B.C., Confucius wrote about a "short-mouthed dog" with short legs and long ears and tails, described as *ha-pa* or "under-the-table" dogs. This would make them quite tiny since tables at that time were only about eight to ten inches off the ground, to accommodate people sitting or kneeling on a mat or cushions in front of it. Some of these dogs were bred to be so small that they could be carried in the wide sleeves of the robes that the nobles wore, thus providing not only companionship, but also additional warmth for the aristocrat on cold winter days. Bronze vessels from the Shang and Chou dynasties show running dogs that look very similar to the Pekingese. However, after Ming Ti associated the little square-mouthed palace dogs with the lion of Buddha, the history of these dogs changed drastically.

In the years that followed, royalty and aristocrats began encouraging the breeding of dogs that looked more and more like the somewhat distorted mental image of a lion that the Chinese now had. During this era, the dogs' long hair (called feathering) increased, especially around the head; the nose widened; and the face flattened. In fact competitions were held every year or so to determine which of the many dogs bred by the servants of the Imperial Palace or in the kennels of the noble houses looked like the best example of the Chinese conception of the Lion of Buddha. The dog that won had its portrait painted and was included in the official records of the Emperor's reign. In addition the person who bred the winning dog received a rank in the civil service that entitled him to a pension for life—obviously a very valuable prize.

Ownership of these royal Lion Dogs eventually became legally restricted to the aristocracy. It was unlawful for Pekingese to be owned by anyone except the nobility or certain high-placed priests. If a commoner encountered one of these dogs when it was crossing the courtyard, the person was required to avert his or her eyes in a respectful manner, much the way commoners did when encountering a noble or highborn person in the street. The dogs were often given titles and honors as prestigious as "Viceroy" by the rulers of the day, and at the very least were treated as if they still had the title of "Imperial Guard," awarded to them by Emperor Ming Ti.

Sometime in the 1500s the Lion Dogs began to disappear from sight, and all breeding of them became confined to the Forbidden City. This is a huge, fortified enclave, located in the middle of Beijing, that covers 720,000 square meters and contains about 800 buildings. The Emperor and his court lived inside the city, and the Pekingese dogs lived out their lives in the royal apartments and kennels, isolated from the rest of the world. The imperial kennel for breeding the Lion Dogs was surrounded by a high-walled courtyard, and a special staff attended to the dogs. All events occurring in the kennel were reported to the chief eunuch who, in turn, would relay information about such things as new litters of puppies to the Emperor.

The first real crisis for the Pekingese came in 1891, at the end of the Boxer Rebellion, when the Forbidden City was invaded by foreign troops and the Emperor Xianfeng fled with all of his court. The power behind the throne, however, was Tzu Hsi (also called Cixî), who would become known in the West as the "Dowager Empress of China." When she organized the evacuation of the palace, she left instructions that all of the dogs that

Queen Victoria named her Pekingese "Looty," since it was part of the loot from the Forbidden City.

had not been taken by the royal party should be killed to prevent their falling into the hands of the "foreign devils." These instructions were not carried out, however, because as chance would have it, an elderly aunt of the emperor remained. When the British troops entered, she committed suicide but she was found surrounded by the last five Pekingese remaining alive in the palace. They were scooped up by a Lieutenant Dunne and taken to England. One pair was given to the Duchess of Wellington, another pair to the Duke of Richmond, and the last dog to Queen Victoria. The Queen named her dog "Looty," since it was part of the loot from the Chinese palace.

In later years, after peace had been established, Tzu Hsi returned to the Forbidden City as empress. As a token of goodwill to the West she presented several dogs of royal breeding to some prominent Americans, who included John Pierpont Morgan and Alice Lee Roosevelt Longworth, the daughter of President Theodore Roosevelt.

In the 1930s, the office of the Chinese Emperor was terminated, as was the protected status of the Pekingese. Now these former royal dogs could become pets for those that could afford to obtain one. However, with the rise of Mao Zedong, and the establishment of the People's Republic of China in 1949, many viewed the owning of dogs, particularly small dogs, as simply a holdover from the country's earlier class-based society—and symbols of status, aristocratic privilege, or rank. The Pekingese and any breeds that had a similar look, such as shih tzus, Lhasa apsos, and pugs, were considered to be particularly offensive to the new communist state. Such dogs became quite rare because their owners were considered "bourgeoisie" and were treated as counterrevolutionary threats to the new state.

In 1966 Mao began the so-called "Cultural Revolution," a declared attempt to continue the class struggle, during which anything associated with privilege or status was attacked by militant hordes of civilians (mostly students and young people), who used violence and mob justice to enforce their views. Pets of any sort, but dogs especially, were reviled as class symbols, and the Red Guards killed them by the tens of thousands. Pet dogs were seized and beaten to death in front of their owners, and any protests by those owners resulted in those people also being beaten (or even killed). Not only pets but also dogs being bred for their meat in Southern China were exterminated, and gourmets were dissuaded from tasting the rich dog flesh, lest they "become infected by class depravity." After nearly ten years of such activity, when the Cultural Revolution finally wound down, the only dogs remaining in the country were mostly working animals (police, military, and herding dogs) and some livestock for meat (mostly chowchows in northern and central China). Unfortunately, very few small companion dogs survived.

In 1977 one British journalist claimed that the only Pekingese dogs that he had found in China were those used as display animals in zoos.

The years since the Cultural Revolution have been kinder to dogs. When as a means of population control, the Chinese government instituted a "one child per family" rule, there was a large public outcry and somewhere along the line, government officials recognized that most humans have a need to nurture. Many parents, after their child was mature and had left home, felt a psychological emptiness. Some government officials discovered that this need to nurture could be reasonably met through caring for and association with companion animals. For this reason the population was once again allowed to keep small pet dogs, and by 1990 several kennel clubs had been established in China.

In some respects this campaign to allow pet dogs has been too successful. At the time of this writing, there are over 660,000 dogs registered in the city of Beijing, and it is believed that there are an equal number of unregistered dogs. Dogs have been proliferating so quickly that the country has had to introduce a "one dog per family" rule to curb the canine population. One stated reason for the restriction on the number of dogs is the fear of rabies, in a country where only about 3 percent of dogs are vaccinated.

The demand for so many companion dogs has caused some interesting reversals in the status of certain dogs. Most of the Pekingese native to China, for example, had disappeared because of social pressure on those who owned this breed. This ultimately caused the Chinese to turn to the West to rebuild their breeding stock.

According to Chen Yin of the China National Kennel Club, "There is now much pride and status for those who own a Pekingese whose ancestry can be traced back to the dogs taken from the Forbidden Palace by the British, or perhaps to those dogs whose forefathers were gifts from Empress Tzu Hsi to important Americans. These are considered to be the true pure examples of the breed—not those hidden away in the countryside for so many years that may have had their bloodlines weakened or polluted by indiscriminate breeding."

So the Lion Dogs of Buddha have returned home. Their bloodlines are still important, and their imperial status is still recognized despite their long sojourn in faraway lands.

Chapter 22

When a Marriage Goes to the Dogs, Who Gets Fido?

IT WAS A contentious divorce: the couple disputed the care and custody of a dependent known in the court papers only as "Baby." According to New York lawyer Patricia Rouse, proceedings nearly ground to a halt when the judge asked for more information about "Baby," only to find that the court was being called upon to settle a custody dispute over a four-year-old German shepherd.

In our modern society, dogs are often viewed as children. We buy them toys, worry about their schooling and manners, and even talk to them in the same singsong tones we use when speaking to babies and young children. Research shows that in

38 percent of divorce proceedings involving dog owners, neither party wanted to give up their four-legged "child." This has caused a legal crisis, since according to laws in most jurisdictions, a dog is merely property and is to be considered much like furniture or real estate. This means that in divorce proceedings, courts are supposed to concern themselves only with the ownership and monetary value of a dog.

The first deviation from that view came in 1942, in a Chicago divorce court. Ruth Schiller traveled 1,500 miles from her new home in Florida to plead her motion for custody of Kiddo, a black cocker spaniel. Her husband asked for the dog as well. Judge Joseph Sabath startled the legal world when he awarded the Schillers joint custody, specifying that each should have the dog for six months of the year.

In 1983, California State Judge John Woolley went even further. After a year of bitter divorce proceedings, Rex and Judi Wheatland were still battling over Runaway, a two-year-old cockapoo. Rex offered $20,000 for Judi's "share" of Runaway, but she refused his offer, saying, "She's my baby. I wouldn't give her up for anything."

During the trial, Mr. Wheatland testified that "Runaway was the nucleus of our family," and to bolster this claim, he brought some of Runaway's toys to court, including a rubber hamburger and rubber hot dog. In addition, he brought a portrait that he had had painted of Runaway.

Judge Woolley granted joint custody to Rex and Judi, which was not a surprise, since this solution was already becoming an accepted practice. What was a surprise was his reasoning. The judge maintained that "*as a child substitute,*" Runaway's case had to be dealt with "in accordance with California's child custody laws."

The treatment of dogs caught up in divorce settlements is becoming ever more like that of children. Thus when Michael Fore and Sheila Mathews of Hennepin County, Minnesota, finalized

The judge said that as a "child substitute," the dog's custody had to be dealt with in accordance with California's child custody laws.

their divorce, they were supposed to share custody of their golden retriever, Rudy. When Fore did not return Rudy after his scheduled visit, Mathews went to court seeking immediate "emergency relief." She went on to justify this by noting in her request: "I do not wish to go any length of time without my pet." The judge did not laugh this request out of court, but rather responded in the same way that he would have to a failure to return a child from a scheduled visitation. He ordered the sheriff's office to send a deputy to "enforce the custody order" and return Rudy to Mathews.

Even the pattern of decisions about custody of dogs seems to mimic the pattern associated with child custody. Just as in the case of human children, judges seem predisposed to grant custody to women, with 81 percent of rulings favoring the former wife. Furthermore, when women have won custody of the dog, their former spouses had been granted visiting rights in a meager 11 percent of the cases. In those rare cases where the man has been granted custody of the dog, however, the ex-wife has been granted visitation rights in 83 percent of the proceedings. And when dependent children have been involved, the custody of the dog usually has been awarded to the parent who gets custody of the children.

Most typically, however, people who get involved in dog custody cases are childless—young couples who have not had children or older couples whose children have grown up and left home. Usually they are reasonably well off, with enough money to spoil their pet with gifts and fancy treats, to dress them up in cute outfits like children, or to buy them fancy accessories— meaning that they also have the available funds to mount a dog custody case. One Vancouver lawyer shook his head and mused about this when we were having lunch one day. He said, "They're paying me $400 an hour to fight over a pet. Think about it. Does it make any sense? Either they are doing it simply to spite their partner, or is it possible that they really do think of their pets as children?"

In many cases, judges seem to be making decisions to award custody of dogs based on the same criteria they would use to determine the custody of children. In Newport Beach, California, for example, a judge awarded custody of two rottweilers, Guinness and Roxi, not on the basis of which person had the more valid ownership claim, but rather according to "what was best for the dogs."

In 2002, in San Diego, Stanley and Linda Perkins disputed the custody of Gigi, a pointer-greyhound cross. They submitted themselves and the dog to "bonding" tests, during which animal behaviorist Dr. Lynn Wilson watched them interact with the dog, looking at which person the dog stayed closest to or followed. The wife's lawyers even presented a specially produced

Day in the Life of Gigi video, which showed the dog playing at the beach, out on a walk, and at rest under Linda's desk. Nearly half of the three-day divorce trial involved arguments over Gigi. In total, legal fees amounted to over $200,000. In the end Linda won full custody of Gigi, and Stanley went to the animal shelter and adopted another pointer cross, named Amy.

The idea of a "bonding" test was picked up by a St. Louis judge in the custody case of a mixed-breed dog. The judge wanted to know which member of the divorcing couple the dog had most strongly bonded with. Both parties were ordered to stand on opposite sides of the court and call the dog at the same time. The plan was to award the dog to the person the dog went to. Instead, the poor dog became so confused that it ended up going to the judge.

In all such cases, a lot of discretion is left to the judges, some of whom still view dogs as property. Thus in 2002, a Pennsylvania appeals court dismissed an application for visiting rights to a dog named Barney because it was "analogous, in law, to a visitation schedule for a table or a lamp."

Some judges are uncomfortable with these contradictory aspects of the law. A dog is the legal equivalent of a sofa, but is also a being with complex emotions and intelligence, capable of forming true emotional bonds with people, and protected by anticruelty laws. Following the letter of the law, some judges feel that imposing joint custody is not justifiable, but they also recognize the feelings of all involved.

Judge Michael Pincus of Maryland was faced with just such a problem when Ethan Assam and Jennifer Kidwell were battling over a keeshond named Sable. They appeared in his court two years after their divorce. As part of the settlement worked out by their lawyers, Kidwell had been granted custody of Sable, while Assam had visitation rights. Kidwell decided that she wanted exclusive possession of Sable and refused to grant Assam access to the dog.

Judge Pincus was not amused and told them, "In divorce cases, a judge can order that any property in dispute be sold and the proceeds divided between the warring couple. That will be the fate of Sable if Ms. Kidwell and Mr. Assam fail to live up to the terms of the divorce settlement in the future." Faced with this prospect, Kidwell and Assam took less than an hour to settle their dispute, with an agreement to share time with Sable.

I wonder what King Solomon would have thought had he seen his own technique of settling child custody modified in this way and applied to the custody of a dog?

Chapter 23

Wildlife and Bloody Murder

NO MATTER HOW large and built-up a city is, wild animals will always coexist with the human population right in the heart of the metropolis. Some, like birds, are visible. Some, like rats and mice, we know are there, although we seldom see them. Other, more exotic animals are also there, especially in large parks or undeveloped areas. For example, I live in Vancouver, Canada, whose metropolitan area contains well over two million people. I also live within a mile or so of a large city park. It is not all that unusual for me to see raccoons crossing the street in the early morning or when I drive home from work. One late night I saw a coyote walking down the middle of a street in our quiet residential area. Coyotes have become a bit of a problem in our city, since they sometimes view free-roaming cats or small dogs in backyards as potential menu items.

On one particular Saturday a different species caught my attention. I was working at home on a statistical problem involving some data from my laboratory at the University of British Columbia. The problem was convoluted and complex, and although I felt I was making progress, the answer to my question was still far from clear. I had been working on the computer since before 6 A.M. and it was now around 1 P.M. and I had developed a bit of a headache, probably from eyestrain and mental concentration. I took a couple of aspirin and, because the day was bright and pleasant, decided to take my dog Wizard out for a walk to allow some time for my head to clear.

At that time I had only one dog, my Cavalier King Charles spaniel, Wiz. My bouncy Cairn terrier "Flint" had recently died, and my handsome flat-coat retriever "Odin" was still in his mother's womb and would not appear on the scene for several weeks. Walking with Wizard was never an athletic event, since he moved no faster than a gentle stroll at any time. On this particular day I was grateful for the easy pace since my mind was still involved with my analytic problem, and our outing gave me time to think. I occasionally glanced at my pretty chestnut-and-white-colored dog as he sniffed his way along, monitoring him just enough to make sure that he had not found something that he might misjudge as a "snack" but otherwise not really paying attention.

Because my mind was elsewhere and Wiz was at the far end of a long, extendable leash, I never saw the actual encounter. I was jolted from my musing by Wizard's yelp of surprise and distress, which was followed seconds later by the unmistakable and unpleasant odor of skunk spray. Wizard dashed back from the bush that he had been sniffing, bringing an intense wave of that noxious scent with him.

I quickly looked him over, to make sure he had no bites or injuries. Although Wiz had been vaccinated, I still worried about rabies, which can be carried by skunks. I also looked at Wizard's eyes, which were large and prominent as in all members of his breed, and fortunately they were clear. Skunk spray does bad things to a dog's eyes and can cause temporary blindness and long-term irritation. If a dog is sprayed in the eyes, a visit to the veterinarian is best, although many people claim the irritation can be handled by flooding the dog's eyes with over-the-counter eye drops, olive oil, or, surprisingly, an over-the-counter vaginal douche. I could see, however, that the faint yellow slick of skunk spray had hit his neck and one side of his body.

I gave a tug on Wizard's leash and started home at as fast a trot as I could coax from him. Fortunately, we were only a few city blocks from my home. I did deviate from the fastest route by crossing the street when a mother with a baby in a stroller approached—there is no sense in exposing a child to the odor of skunks at that early age. Attracted by the sight of my trying to trot with a resistant and distressed dog, and also puzzled by the stench that hovered over us, several neighborhood kids approached.

"Don't touch him," I warned them. "Skunk spray is an oil that spreads really easily and is difficult to get off. I just have to get him home and cleaned quickly. If you leave the spray on too long, it gets into the hair and skin and then you can have it stinking up the place for months."

The kids backed off at my warning, but followed me out of curiosity.

Once home, rather than going inside, I rang the doorbell. My wife, Joan, appeared a couple of moments later and after one sniff knew what the problem was.

"Don't bring him into the house," she said with a sound of disgust in her voice. "There is tomato juice in the cupboard."

I handed her the leash and dashed inside. There, I quickly

changed into an old shirt and ripped jeans I wouldn't mind losing if the smell couldn't be washed out, then grabbed the large
can of tomato juice and some liquid dishwashing detergent. I
returned to the door and took the leash and Wiz. We walked
down the steps and around the house to the yard with a small
entourage of neighborhood children of various ages. By the time
I got to the rear, Joan had reappeared with a yellow plastic wading pool she had gotten for our small grandchildren to play in.
She was also trailing the garden hose behind her.

I put on some rubber gloves and slipped Wiz into the plastic-
walled pool. We had always been told that the way to deal with
skunk smell on dogs is to soak them in tomato juice. In any
event, that was the only remedy that either my wife or I could
think of at the moment, so I began to splash my unhappy dog
with the contents of the can. The kids watching thought this was
hilarious and giggled while I poured the heavy red liquid over
him. Although Wizard was obviously distraught, he eventually

Some neighborhood kids gathered to watch me pour tomato juice
over the victim of the skunk attack.

accepted the situation and after a while lay down passively on his side. His only movements from then on were an occasional glance up at me with an expression that seemed to ask if this day could possibly get any worse.

While I waited for the tomato juice to soak in and do its work, one of the neighborhood kids produced a camera and ran around snapping a few photos. After about ten minutes of soaking, I started the work of washing Wiz down with the liquid detergent. The pink-colored suds looked a lot more comforting than the pools of bloodlike tomato juice.

Since Wizard's encounter with the skunk, I have learned a number of alternatives that are less messy than tomato juice, including white vinegar and apple cider, although there is some question as to how well these home remedies work. Commercial products that also remove the skunk odor are available, but most people don't keep containers of them around on the off-chance of a smelly encounter.

There is one home remedy that really does the job quite well. It was developed by Illinois chemist Paul Krebaum while he was working on ways to scrub noxious odors from industrial waste-gases. The recipe is simple: In an open plastic bucket, start with 1 quart of 3 percent hydrogen peroxide. It is important that it be fresh, from a never-opened bottle, and since it is usually sold in pint containers, two are needed. Next, mix in ¼ cup of baking soda (sodium bicarbonate)—baking powder won't work. Finally, add 1 to 2 teaspoons of liquid soap or dishwashing detergent. For large dogs, add up to one quart of lukewarm tap water to have enough liquid to cover the dog completely. Hydrogen peroxide and baking soda, when combined, become a sort of chemical engine for churning out oxygen, so there will be some foaming, which means that you have to use the mixture immediately after mixing. The soap breaks up the oils in the skunk spray, allowing the other ingredients to do their work. Don't try to store this mixture in a closed container, since the oxygen buildup may cause it to explode and, besides, it won't work if it isn't fresh.

Wash the dog promptly with this mixture (avoiding getting anything into his eyes). Thoroughly work the solution deep into the fur. You should leave the solution on for about five minutes or until the odor is gone—let your nose be your guide here. Some heavily sprayed areas may require a "rinse and repeat" washing.

Had I known about this remedy at the time, the complications that followed could have been avoided: Once the kids had wandered off and Wiz had dried off, I anointed him with a couple of drops of vanilla to mask any residual odor, then returned to my normal activities. Perhaps it was the surge of adrenaline or the physical activity, however, once I was back at the computer, my previous problem seemed to solve itself, so I quietly thanked the skunk for his contribution. What I could not know was that the faint click of the camera shutter during Wiz's tomato juice bath would echo loudly over the next couple of weeks.

One of the teachers in a nearby school had hit upon an interesting way to encourage her students to do more writing. Rather than having them produce traditional essays, she had each of the kids create an individual website. This was so that each student could maintain a *blog*, which is short for "web log." These blogs were supposed to be a kind of Web diary, where students would write about their hobbies, likes and dislikes, family activities, current events, or simply amusing or interesting items they had encountered. Each student was required to make a minimum of three entries per week, and the teacher would monitor, grade and critique their entries over the semester.

One of the students in this class happened to be the kid with the camera during Wizard's skunk decontamination bath. In an attempt at humor he posted one of the photographs he had taken as part of an article with the headline, "A Horrifying Case of

Animal Abuse?" The text of the article went on to say that a well-known psychologist, supposedly an expert in dog behavior and a friend of canines, was caught on camera engaging in bizarre activities he claimed were designed to relieve his dog of a noxious problem. Furthermore, he did so in front of an astonished crowd of neighbors. The article finished with the photograph of Wizard lying forlornly in his yellow plastic tub, his fur covered in red liquid streaks, and a giant pool of red fluid around him. It looked for all the world as though he had been killed by some deranged beast that left him lying in a puddle of his own blood. The final sentence of the piece finished with, "Can this be considered some kind of useful treatment or is this bloody murder? Ask the Doctor."

The student sent me an e-mail linked to his Web posting. I looked at it and thought that the picture did in fact look like a gruesome killing. I felt that the humor was a bit heavy-handed and wondered why at no point in the piece did he try to clue the reader in to the fact that the red liquid was tomato juice being used to rid a dog of skunk odor. However, as an educator myself, I noticed that the vocabulary and sentence structure were good, and it appeared to me that the teacher's plan to improve her students' writing was working. In the end I passed it off as just another of life's happenings.

A few days later I received a phone call at my office in the university from the teacher who was having the students write the blogs. She wanted to know if I had seen the posting, and she appeared to be distressed about it.

"I was quite upset to see the evidence of animal abuse that my student collected. Not many psychologists in this area other than you work with dogs, and I need to know if this event involved you or someone else that you may be acquainted with."

"Didn't you ask the student to explain the circumstances around that photo and how he came to get it?" I replied.

"He told me that he felt that by posting material on his blog he was in effect a journalist and had the right to 'protect his sources.' I couldn't convince him otherwise, so I have reported the matter and forwarded his posting and the picture to the SPCA and the police."

I had a sinking feeling in my stomach. If I admitted that the posted photo was of my dog, at the very least the SPCA and the police would feel it necessary to follow up on the incident report by questioning me. If it got out that I was being interrogated because of a charge of animal abuse, this could be both embarrassing and damaging to my reputation, since many people consider the simple fact that a name is mentioned in association with a criminal event as the equivalent of that person being convicted, regardless of later proof to the contrary. I did not want to lie to the teacher, but I felt that the problem could be resolved without involving the authorities or potentially the press.

So I carefully replied, "I certainly did not engage in animal abuse, nor do I know of any other psychologists in this area who have done so."

She accepted my statement and my promise that I would try to clarify the matter, but was still upset and angry when she hung up the telephone. I immediately e-mailed the student and told him what happened. Rather than be harsh and threatening, I simply suggested that it might make an interesting second posting on his blog to let his readers in on the joke. Fortunately, he agreed, and that same day additional photos showing my hand pouring tomato juice from the can onto Wizard, followed by the later detergent scrub with the pink foam, appeared on his website, along with an explanation that the incident was in response to a skunk attack. He must have cropped the photos, since my head and face were never visible, and he had the journalistic reserve to leave my name out too. I assume that the teacher accepted this entry because I never heard from her again and was not questioned by any official agencies about this matter. Unfortunately, some other people to whom I told the story alerted me that the original photo had been picked up by some animal rights groups and presented on their websites as "evidence" of the fact that psychologists are abusing and killing animals as part of their research.

After the new photos were posted, I sat nursing a large glass of bourbon while I explained the day's events to my wife. Joan looked at me and said, "You didn't have to be so devious, you know. Why didn't you just invite the teacher over to show her that Wizard is alive and well?"

I looked down at my dearly-loved dog and shook my head. "I didn't want to have to explain or deal with the side effects of my tomato-juice deodorant treatment," I said.

"What side effects?" was her puzzled response.

"Look at him, Joannie, all of his lovely white fur is now bright pink and I don't have a clue as to how long the color will last!"

Fortunately, a week and a half later, after three baths with a "whitening shampoo" for dogs, all of the pink evidence of my episode of "animal abuse" had been erased from Wizard's fur. Unfortunately, it was not erased from the Internet, since I recently ran across that photo of Wizard's tomato-juice bath still posted on an anti-animal cruelty website as evidence of vivisection and animal torture being done by psychologists.

Chapter 24

Astromutts

FEW NORTH AMERICANS remember that the first living Earthling to orbit the earth had four legs and a tail and answered to the name of Laika. We are so accustomed to the idea of men in space stations and even walking on the moon that it is hard to believe that there was a time when there were grave doubts about whether humans could survive outside of the Earth's atmosphere. Among a long list of potential problems that scientists were concerned about was the harmful radiation normally filtered out by more than 50 miles (80 km) of atmosphere. It was also feared that the absence of gravity might wreak havoc on the vestibular system, which is responsible for our sense of balance, and that the resulting nausea and vomiting might completely incapacitate astronauts. Scientists further worried that the stresses of being launched into space—the high gravity forces and extreme vibration—might be harmful. There were other concerns as well, including the simple fear of the unknown that kept space scientists awake at night.

After World War II, the exploration of space became a worldwide passion and a great political issue. America began its space program by using the expertise of former German scientists and captured German rocket technology, such as the V2 rocket, which had originally been designed to be a ballistic missile and carry explosives. However, the Soviet Union had also captured German scientists and had the same technology. Both the United States and the USSR were striving to be accepted as the greatest superpower in the world, and part of this goal involved demonstrating their advanced scientific and technological accomplishments. It soon became clear that the symbolic winner would be the first nation to conquer space, to place a man in orbit, and to walk on the moon. The great "space race" was now in process, with the reputations of both nations and their national pride hanging in the balance.

Because of fears that space might be lethal to people, the first living organisms launched into space were fruit flies and kernels of corn. These took a suborbital jaunt on an American V2 rocket in 1947. However, on July 22, 1951, the Soviets astounded the world when they sent two dogs, Dezik and Tsyganka, up to the edge of the atmosphere—a distance of around 100 kilometers—then had them leave the rocket in ejection seats and float safely back to earth. Both dogs returned healthy and unharmed. Tsyganka (Russian for *Gypsy*) was retired from the program and adopted by the Soviet space scientist Anatoli Blagonravov, who later would become well known for his efforts to achieve peaceful international cooperation in space research. Dezik was not so lucky. Some two months later, she took another suborbital flight with a dog named Lisa, and neither survived.

The Soviets eventually used twenty-one dogs in suborbital

flights, reaching heights of nearly 500 kilometers. The Russians next shocked their U.S. counterparts by placing in Earth orbit the first satellite, *Sputnik 1*. While the world was still buzzing about that accomplishment, the Soviets topped that triumph by putting the first living creature from Earth into orbit. *Sputnik 2*, launched on November 3, 1957, carried a dog—the Samoyed/terrier mixed breed, Laika.

Dogs were chosen for these experimental flights into space for a number of reasons. Dogs, for example, are intelligent, adaptable, and well suited to endure long periods of inactivity. Their physiology is closer to that of humans than many other species used in research, including rats. Also, the Russians felt that there had to be lots of publicity around each new accomplishment in order to convince the world that they had the most technologically-advanced space program in the world. A dog is a more impressive spacecraft passenger than rats or mice. Furthermore, they can be placed in front of television cameras and bark into microphones, which was what Laika was called upon to do prior to her flight. Laika's publicity activities and the timing of her mission were personally orchestrated by then-Prime Minister Nikita S. Khrushchev, who wanted her flight into space to be part of a gala celebration for the fortieth anniversary of the Bolshevik Revolution.

Only mixed-breed dogs were used in the flight program, because the Russian scientists felt that purebred dogs might suffer from genetically-related health problems. All of the dogs in the program were actually stray dogs, rescued from the streets of Moscow, rather than animals accustomed to living in a house. This was preferred because the researchers felt that these dogs would be able to tolerate the rigors and extreme stresses of space flight better than animals that had lived a more comfortable life. Only female dogs were used, since they did not have to lift their leg to urinate, and it was easier to design equipment to collect urine and feces from them.

Like the human cosmonauts who would follow them, all dogs in the space program were trained for the rigors of orbital flight. As part of their training, they were confined in progressively smaller boxlike containers, and the confinement times were increased until durations of about three weeks were reached. They were trained to remain motionless for long periods of time while wearing spacesuits, and were placed in simulators that provided noise and shaking similar to that of a rocket during takeoff. In addition they were given rides in centrifuges to simulate the increased gravitational force associated with a high acceleration rocket launch. They also learned to eat from a dispenser that provided them with a nutritious jellylike protein and to drink from a special fluid dispenser.

Laika had some additional training, since hers was an image-building political mission as well as a scientific one. So she had to learn her new name Laika (meaning *Barker*) because the foreign press was having difficulty pronouncing her real name Kudryavka (*Little Curly*).

Laika's trip was always meant to be one-way. According to Russian reports at the time, she survived for four days in good health and was euthanized by a poison in her last portion of food so that she would not suffer. In 2002, forty-five years after her flight, the truth came out when the Russians admitted that there was a malfunction in the spacecraft and Laika died after only five hours in orbit, presumably due to overheating, fear, and stress. The false reports were part of the publicity aspect of the flight. It was an attempt to minimize the unexpectedly bad press the Soviets were getting for deliberately sending a dog out to die in space. Oleg Gazenko, one of the scientists responsible for Laika's death, has said, "The more time passes, the more I'm sorry about it . . . We did not learn enough from this mission to justify the death of the dog."

A space physiologist who worked with NASA once explained to me why the Americans never used dogs in their space program. "We told the public that we were using monkeys because they were closer to humans in their physiology and behavior. We even broadcasted films of Ham, a chimp, pulling levers for treats in a spacecraft in 1961. We said that this demonstrated the ability to perform tasks in space, which was why we chose primates. Actually, we could have had a dog doing tasks in space just as well. The real reason that we didn't use dogs was because of the kind of public response that resulted when Laika died in orbit. There was real public outrage over the death of that dog. People don't like it when a monkey is killed in an experiment, but nobody lives with a monkey, so their response is kind of bland and academic. When it comes to dogs, however, there is a real emotional response. People can imagine the 'victim' being their own pet. We needed public support, and letting a dog die in space seemed like a sure way to damage NASA's reputation."

The Russians did manage to pursue their advantage in the space race and at the same time regain some favor in the hearts of the people when *Sputnik 5* was launched on August 19, 1960. The passengers this time were two dogs, Belka (*Squirrel*) and Strelka (*Little Arrow*), plus forty mice, two rats, some flies, and a selection of plants. These animals spent a full day in orbit, and then the spacecraft's retrorocket was fired. The landing capsule and the dogs were safely and speedily recovered and the passengers became the first Earthborn living animals to survive orbital flight. To the delight of the world, the dogs were paraded in front of television cameras and shown to be happy dogs in obvious good health.

Strelka would provide further good press for the Soviets. In June 1961, a summit meeting took place between then-President John F. Kennedy and Soviet Prime Minister Nikita Khrushchev in Vienna. At dinner one evening, Khrushchev was seated next to the president's wife, Jacqueline, and he was bragging about the flight of Strelka and Belka. He went on to announce that Strelka had just given birth to six healthy puppies, proving that

Strelka (left) and Belka, the first animals to go into orbit and to safely return to Earth.

space flight was safe. Mrs. Kennedy laughed and said, "Oh, I bet that our daughter Caroline would love to have one of those pups." Two months later, Soviet Ambassador Mikhail Menshikov showed up at the White House with a little white puppy named Pushinka (*Fluffy*) from Strelka's litter.

When Pushinka arrived, the Kennedy children immediately wanted to play with her. The Cold War mentality of that time, however, prevented that from happening at first. The Secret Service was certain that dog had been implanted with some sort of transmitting device to spy on the president's activities, or even worse, some sort of lethal biological implant that might ultimately spread infection through the First Family and the White House staff. So they turned the dog over to the CIA, whose staff were going to kill and autopsy the dog. The president was horrified. Killing a dog with such a famous mother, which had been given as a peace offering from the Russians to his children, would have been a public relations disaster. So, instead, the dog was sent to Walter Reed Army Medical Center for examination. She was poked, prodded, scanned with a magnetometer, tested with thermography, X-rayed, and even subjected to an early version of a sonogram. All the CIA found inside of Pushinka was the usual "doggy stuff," so she was returned to the White House. Around a year later she gave birth to a litter of what the president called "pupniks," fathered by Kennedy's Welsh terrier, Charlie.

The last Russian dogs to fly in space were Veterok (*Breeze*) and Ugolyok (*Little Piece of Coal*). They were passengers on the biosatellite *Cosmos 110*, which was launched on February 22, 1966. The dogs were observed in orbit for twenty-two days via video transmissions and biomedical telemetry. The spacecraft then returned to Earth safely with the dogs in good health. This still stands as the canine spaceflight duration record and was not surpassed by humans until the flight of *Skylab 2* in June 1973.

After *Cosmos 110*, dogs were removed from flight status in the Russian space program. The reason given to the media was that enough had been learned, and it had been established that mammals can live and work safely in space conditions for long durations. The other reason, however, was that of the thirteen dogs that had been sent into space in the Soviet orbital program, five had died. Each dog's death had brought a storm of resentment toward Russia and its space efforts, and the government no longer wanted the bad publicity.

There was one last publicity effort to minimize any damage to Russia's reputation caused by the use and death of dogs in that nation's space program. It involved establishing a permanent exhibit celebrating the "canine cosmonauts" in the Memorial Museum of Space Exploration in Moscow. The bodies of Belka and Strelka were preserved using skillful taxidermy, and Belka can be seen in a glass case at the museum. Strelka is often on tour as part of the museum's traveling exhibition. The names of all the dogs that flew in the Soviet space program are listed as part of the display, and the heading proudly designates them as "national heroes." It appears that average Russian citizens not only had strong sentiments about the use of dogs in space but also continue to have a real sense of pride in those that survived.

Chapter 25

Semper Fido

ON FEBRUARY 21, 2000, a monument was dedicated to honor a group of heroic soldiers. In the long tradition of fighting troops, they had odd nicknames, such as Duke, Pepper, Cappy, Sparky, Gunner, and the like. The fact that these soldiers were dogs, not men or women, really does not matter. They had shown courage, loyalty, and dedication to duty, saving the lives of many of their comrades, and many had given their own lives to preserve freedom. The Allies used 30,000 dogs in World War I, while the United States alone used 12,000 dogs during World War II, 1,500 in Korea, and another 4,000 in Vietnam. In the first Gulf War in Iraq, the allies employed around 1,600 dogs, and in current military actions, about 2,000 military working dogs are facing danger daily alongside American soldiers, serving mostly in Iraq.

The abilities of war dogs are well documented. With their superior senses they can often detect enemy soldiers hidden in ambush at a distance of up to 1,000 yards. Dogs can determine the scent of buried land mines that metal detectors would miss because the devices are made of plastic or ceramics. They can even smell the breath of saboteurs trying to elude detection by approaching underwater, using hollow reeds to breathe through.

The efficiency of these dogs is remarkable. During the war in Vietnam, Captain Richard Hale ran the dog training camp at Bien Hoa. There, the dogs learned to detect booby traps, which were one of the major sources of American casualties. Hale has said that even when the Viet Cong attempted to disguise the appearance and scent of the area around such traps, the dogs' hearing was so good that they actually could hear the faint, high-pitched sound of a gentle breeze passing over the trip wires. That sound was sufficient to prompt them to give the "alert" signal. Hale has reported that one of the dogs detected 159 booby traps in one year of service. "That is 159 legs or lives saved right there," Hale said.

The personal heroism of military dogs is also recognized, although less well known by most people. One of the most famous dog heroes of World War II was Chips, a German Shepherd-collie-husky mix that worked as a tank guard in General George S. Patton's Seventh Army. Chips's duties took him through eight combat campaigns in Africa, the Mediterranean, and Europe. He was even selected to be a sentry at the conference between President Franklin Delano Roosevelt and British Prime Minister Winston Churchill in 1943. His true courage was actually tested on the beaches during the Allied invasion of Sicily, when American troops moved up the beach, nearing what they thought was an abandoned enemy pillbox. Suddenly, Chips bolted forward and charged into the emplacement. Next, his fellow soldiers heard a commotion, some shouting, and a shot. When they rushed to investigate, they found that the emplacement actually held six Germans who were preparing to open fire on the passing troops with a machine gun. Chips had subdued the gunner, and the other five were cowering against the far wall, ready to surrender to the approaching troops. Chips had accomplished this feat despite having been shot as he began his attack. For this heroism, he received a Silver Star and a Purple Heart directly from General Dwight Eisenhower.

The bond between a military dog and his handler runs deep. In Vietnam, near Da Nang, for example, a soldier, John Flannelly, and his German Shepherd scout dog Bruiser were on patrol. Bruiser was acting cautiously, and Flannelly moved silently with him, searching for any sign that the enemy might be near. Suddenly Bruiser stopped moving, and his ears began to twitch— a signal for danger. Then he marked a specific location by staring at a dense section of brush and trees that could be the perfect setting for an enemy ambush. Based on the dog's behavior, Flannelly signaled for the squad to fire, and seconds later, a hail of enemy bullets and grenades erupted from the very place that Bruiser had pointed out.

During the action, Flannelly was severely wounded. Fearing that he would not survive, and worrying about the dog's safety, he tried to get Bruiser out of the line of fire by ordering him to retreat to the rear. For once, Bruiser ignored his handler's commands and instead started to drag Flannelly back toward cover. Suddenly the dog let out a yelp and blood spurted from his flank. Ignoring the bullet wound, he continued to drag his human comrade toward safety, but then again the big dog let out a cry, and another geyser of blood erupted as he took a second hit. Bleeding profusely, he once more returned to his master and somehow managed to drag him out of the line of fire and back to his comrades before collapsing in a blood-soaked heap. Several soldiers did die that day, yet because of Bruiser's advance warning, many survived, including Flannelly. At the field hospital, where they bandaged him and gave him painkillers in preparation for transfer off the front lines, the despondent soldier sadly thought of the dog that had fallen in the act of saving him. It was then that a stretcher was moved next to his. A medic looked at Flannelly and pointed to the furry, bandage-swathed occupant and said, "It looks like both of you are going to make it."

Flannelly would later say that "Bruiser saved not only my life, but the lives of the other marines I was working with. I never would have made it without him."

Perhaps the most distressing aspect of the story of the dogs of war is the way in which they have been treated in recent years by the military bureaucracy. At the outbreak of World War II, a civilian organization called "Dogs for Defense" headed a campaign calling upon patriotic citizens to volunteer their dogs to serve in the military. The people who allowed their pets to be recruited were promised that the dogs would be returned to them at the end of the war. However, as the war drew to a close, the government reneged on its promise. While the troops were boarding ships to return home, many of these valiant dogs were callously euthanized.

Marine Captain William Putney commanded the Third War Dog Platoon, which served a vital role in the liberation of Guam. When Guam was declared "secured," more than 8,000 Japanese hidden in the dense vegetation of the island refused to surrender. They were excellent jungle fighters and could set up deadly ambushes, with virtually no chance of human detection until it was too late. It was only through the alertness of the scout dogs that the hidden enemies were located and the island eventually made safe.

Putney was horrified when he learned that the canine heroes that he had served beside were being summarily killed. The military tried to justify these actions by claiming that the dogs might be carrying diseases and that they were psychologically unfit to be returned to civilian life. Putney had worked with these dogs and knew that it was because of their love for their handlers that they had fought and served so well. Few war dogs are naturally aggressive, and most must be trained to attack people. For these reasons he believed that military dogs could be retrained to return safely to their homes. Putney and others threatened to make a public issue of the military's betrayal of their trust to the dog owners who had contributed their pets to the war. Because of this, a reluctant military established a rehabilitation program for war dogs.

Putney was correct. Although many dogs had died in action, and many others already had been killed as part of the initial bureaucratic decision, there were still 559 dogs enlisted in the Marine Corps. These were sent to Camp LeJeune for resocialization. Of these, 540 were retrained and released to civilian life. Of the nineteen that had to be destroyed, fifteen were euthanized due to health reasons, while only four had incorrigible behavior and had to be put down. Thus 99 percent of the war dogs were psychologically fit to be returned for a well-deserved retirement in a loving home.

Retraining recruited service dogs is expensive. For this reason the military changed its policies so that no "volunteers" would be taken and dogs would simply be purchased from breeders. In addition the dogs were reclassified and labeled "equipment," which made them exploitable and disposable. Accordingly war dogs lost their rights to be considered soldiers, to earn a rank, and to receive medals for their bravery. In fact, they also lost their classification as living creatures.

Because of this policy change, of the 4,000 dogs sent to Vietnam, minus the 500 that died in combat, only 204 of the remaining dogs ever came home again. In March 1973, when the United States announced its withdrawal, military authorities handed over all of their "surplus equipment" to their South

Vietnamese allies. This included jeeps, guns, bullets, and all of the scout and sentry dogs. Evidence shows that many of these loyal, intelligent, and hardworking dogs were later killed and eaten by the Vietnamese soldiers.

One of the people at the dedication of the war dog memorial was Steve Janke. In 1971 he was a nineteen-year-old military policeman who served with a German Shepherd named Kobuc. They were patrolling an Air Force base at Cam Ranh Bay when the dog tensed. Kobuc's keen senses had detected an enemy, even though Janke could see none. He called for help, which turned out to be just in time—North Vietnamese soldiers had infiltrated the area next to the base's fuel depot. Janke survived that firefight, and this would not be the last time that Kobuc would save his life.

Janke stood watching the dedication of the bronze and granite monument. It depicts a soldier in combat gear, designed to represent handlers from all wars, and beside him, alertly scanning for enemies is his dog. In a low voice Janke spoke about all of the dogs like his Kobuc "... they kept you alive when there was no other reason for you to survive."

Janke's eyes took on a wet glaze as he recalled, "I remember the last night that I went to say good-bye to him, I just talked to him like a person. I told him, 'I just want to thank you for saving me, for keeping my parents' youngest son alive all these months.'" He cleared his throat and looked at the ground, "I probably told him things I've never told any human being."

The dogs that served in Operation Desert Storm during the Gulf War did at least come home. Of course there were no honors or medals because they were still merely "equipment." The U.S. Armed Forces publication *Stars and Stripes* tells of one such dog, a Belgian Malinois named Carlo, and his handler, Staff Sergeant Christopher Batta. Carlo was on demolitions detection duty in Kuwait. He alerted Batta to 167 locations where explosives were concealed as booby traps left by Iraqi soldiers (an average of three potentially life-threatening situations per day). In one instance he uncovered cluster bombs hidden under food supplies ready for distribution in a neighborhood where children were playing. Batta received the Bronze Star for his service in Kuwait, while Carlo received nothing. At the close of the ceremony, Batta removed the medal from his uniform and pinned it on his dog's collar, explaining, "Carlo worked harder than me. He was always in front of me."

Given that war dogs saved an estimated 10,000 lives in the Vietnam War alone, it is not surprising that pressure grew to do something to honor these canine soldiers. At the very least they should not be discarded at the end of their service like a broken-down jeep. In the face of a campaign by many veterans' groups and the public-at-large, the U.S. Congress passed a bill in 2000 allowing retired military dogs to be adopted by their former handlers, law-enforcement agencies, or people with adequate experience to handle a dog with military training. The previous ban on allowing such dogs to be released to civilian life had been based on the perception of the possible danger they might pose to the public. The law resolved this concern by including a "Hold Harmless Agreement," which released the United States from any liability for a retired military dog's actions once the dog is transferred to a new guardian. In fact there have been virtually no reports of any problems with these dogs since the program began. Much more common are stories like that of Technical Sergeant Jamie Dana who, in 2005, nearly died in an Iraq car bombing that also injured her military dog, a German shepherd named Rex. Dana's wounds ended her military career. When she began to recover from her injuries, she filed paperwork with Lackland Air Force Base in Texas, which serves as a clearing-house for requests for all retired military dogs. As a result, Rex now lives on a farm in Smethport, Pennsylvania, with Dana. She

The War Dog Memorial at Holmdel, New Jersey.

claims that he acts for all the world as if he was never really meant for a soldier's life. "He loves everybody and he sleeps beside my bed," she says.

Another way to honor these courageous dogs' service has been to erect memorials. Several have been built, financed totally by private individuals and veterans groups. The first two were dedicated at the March Air Force Base in California and at the Infantry School, in Fort Benning, Georgia. Additional war dog memorials have been erected in New York, Pennsylvania, Texas, New Jersey, and Illinois, to name a few. However, it was not easy to find a place for these memorials because there are still some people who fail to recognize the role that dogs have played in military actions. The one at Fort Benning was originally scheduled for the Riverside National Cemetery; however, authorities refused to allow it because some veterans groups suggested that a memorial for dogs at the cemetery would be disrespectful to the service personnel who had died in combat.

John Burnam, a veteran who has helped raise funds for the war dog memorials, understands these sentiments. He spent countless days during the Vietnam War with his German shepherd, Clipper, and argues that this dog saved his life and the lives of his comrades on patrol with him several times. Says Burnam, "We aren't equating them to humans, but we are saying . . . there are families that have grandkids as a result of these dogs being deployed."

At the base of the National War Dog Memorial is an inscription which probably reflects the sentiments of all of the men and women who served beside loyal and valiant dogs in time of war. "They protected us on the field of battle. They watch over our eternal rest. We are grateful."

Benefits of Dog Ownership

Chapter 26

The Curse of the Vampire

"I DON'T UNDERSTAND why you are willing to expose your children to life-threatening health problems because of that thing!" my Aunt Sylvia said in her usual authoritative manner. Her eyes were on my mother, but her finger was pointing at my beagle, Skipper. The little dog recognized the threat in Sylvia's gesture and loud voice, so he cringed behind my mother's legs looking for safety, all the while casting his eyes about to see if he could determine what he had done wrong to trigger this outburst.

Sylvia was my mother's sister. She had steel-gray hair that she wore tightly pulled back. Intelligent and well-read in a nonacademic and nonscholarly way, she was quick to form opinions and so confident in the correctness of her conclusions that she attempted to impose her views upon members of her extended family as if they were edicts from the gods. Matters involving health, manners, interpersonal proprieties (including discovered or suspected indiscretions), and political issues were her usual topics. That day Sylvia had started on a rant about health and dogs. My mother had learned that it was best to not argue with her but simply to wait until the flow of words slowed sufficiently to allow mild comment, and then to offer her sister a cup of coffee and some sweets as if nothing had transpired.

Skipper's transgression had been to leave muddy footprints on the kitchen floor after he returned from a brief session in the yard. To my fastidiously neat aunt, each paw print was a sign of potential disease tracked in from the dangerous world outside.

"Don't you know the dangers of rabies? You already nearly lost your son to that. How about the deadly diseases that the intestinal parasites in this dog can transmit to the kids? And if you avoid all of that, there is the likelihood that your children will develop all sorts of allergies and sensitivities by being exposed to that dog's hair, his dander, and even the mites and fleas that I bet he carries around with him."

Sylvia had used this well-practiced set of arguments for years to explain to her own children why they could not have a dog. And these beliefs were not unique to her but even today are shared by many people who worry about "zoonoses"—diseases that can be transmitted from animals to humans.

Probably the deadliest and most feared of the zoonotic diseases is rabies. The name comes from the Latin word meaning "madness," "rage," or "fury" because animals with rabies are usually extraordinarily aggressive and may attack without provocation. Many people think of rabies as exclusively transmitted by dogs, but virtually any warm-blooded animal can carry the disease. In developed countries, the majority of dogs are vaccinated against rabies, and the most common carriers are cats, foxes, bats, skunks, raccoons, and rabbits. In human beings the first symptoms appear to be much like that of the flu, with fever, vomiting, and headache. However, problems quickly escalate to partial paralysis along with the production of large quantities of saliva and tears and an inability to speak or swallow, which results in "hydrophobia," where the victim has such difficulty swallowing that he actually panics when given liquids to drink. In addition there is a massive psychological deterioration, marked by confusion, anxiety, insomnia, agitation, paranoia, hallucinations, bouts of hostility, and even delirium. Once the symptoms show themselves, death is 100 percent certain within two to ten days, and death by rabies is quite ugly and painful.

Before 1885, the year that Louis Pasteur and Émile Roux first successfully treated a human rabies victim, the most common treatment for human rabies was euthanasia. Doctors or close family members resorted to smothering the patient with a pillow, which was considered to be much kinder than allowing him to suffer the agonizing death that resulted from the disease. Today, if promptly treated with a series of injections started within 48 hours after exposure, before the symptoms show, rabies in humans is seldom fatal.

Historically, rabies was one of the reasons that dogs were considered to be unclean animals by a number of faiths, including Islam and early Orthodox Judaism. Some scholars suggest that the vampire myth may have been inspired by the reality of a severe rabies epidemic in Hungary that raged from 1721 to 1728. In the Hungarian outbreak, rabies was being transmitted by the bite of infected animals—the bats, wolves, and dogs that are traditionally associated with vampire legends—and also by the bite or simply contact with the saliva of infected people. One side effect of rabies is an extreme sensitivity to light. Because of this sensitivity, victims may refuse to look at their own reflections in mirrors, probably to avoid seeing unexpected, painfully bright patches of light that might be reflected from someplace behind them. Rabies victims tend to avoid water (holy or not), and strong odors such as garlic seem to nauseate them. Confronted with such irritants, a person with rabies may experience an involuntary spasm of the facial muscles, causing him to bare his teeth. He may emit frothing bloody fluid from his mouth, making him look like a ravenous beast that has been sucking blood. Rabies also changes the normal sleep-wakefulness cycle and may cause victims to prowl around at night. All of these aspects of the disease are characteristics consistent with superstitions about vampire behavior.

Part of Aunt Sylvia's fear of rabies had to do with me. When I was about five years of age, she and my Uncle Alex rented a house for the summer in Atlantic City and invited me to spend a weekend there with my cousins Helma, Eleanor, and Steve. Only one day into my holiday, I was bitten by a rabid dog. Treatment was started immediately, beginning with washing of the wound, a cleaning with alcohol, and an application of iodine followed by a series of shots. At that time, the shots were excruciatingly painful. They were given with a wide-bore needle (which looked like a lance to my young eyes) and injected directly into the abdominal muscles. Everybody in the family was in a state of panic; however, after the fifth shot, since I wasn't showing any symptoms, it became clear that I was going to survive. Given the pain involved in that treatment process, some people have wondered why I was not left with a lifelong fear of dogs. The reason is simple: the dog hurt me a little bit and just once, while the doctors hurt me a lot, and many times. So I have been left with a lifelong discomfort associated with hospitals, not dogs.

Today, rabies treatment is much less painful, with the shots administered in locations near the bite and with a normal-sized hypodermic needle. Pre-symptom treatment is virtually always successful. The likelihood of contracting rabies from dogs has been greatly reduced due to effective antirabies vaccines. At the time of this writing, the number of deaths from the disease in the United States is now around two per year, and about the same number annually in the European Union. This means that you are forty-five times more likely to die by being struck by light-ning than from rabies. Typically, rabies-related deaths are from skunk, fox, or bat bites that were not immediately treated.

In some parts of the world, however, data show that rabies is still a major health concern. The World Health Organization reports about 55,000 deaths from rabies each year, the largest number (around 31,000) is in Asia, with India accounting for the most of these (30,000). Africa is next, with 23,000 rabies deaths. In Asia and Africa unvaccinated dogs can be blamed for about 40 percent of these cases.

Another health problem that can be transmitted by dogs to humans is intestinal parasites, of which there are three forms: roundworms, hookworms, and tapeworms. Roundworms usu-ally only affect very young dogs, perhaps up to about 8 months of age. The dog's symptoms include poor growth, weight loss, a pot-bellied appearance, coughing, occasional vomiting, and slimy diarrhea, with worms sometimes visible in the stool. In humans the symptoms are much like a severe influenza: fever, muscle pain, coughing, and loss of appetite. The second form of parasite, hookworm, has symptoms that show up in dogs as anemia or bloody stools, weight loss, loss of appetite, poor growth, coughing, and skin inflammations, especially on the feet, legs, and abdomen. In humans the problem shows up as localized skin inflammations, especially in the lower leg region.

Since both of these problems are passed to both dogs and humans through contact with the stool of infected animals, pre-vention is simple: clean up after your own dog has had a bowel movement, and make sure that your dog does not come into contact with the droppings of infected dogs. Safe, effective de-worming medications are available, many of them inexpen-sive, over-the-counter products. Finally, the simple expedient of washing your hands before eating if you have been playing with the dog prevents the problem.

Dogs can become infected by tapeworms through exposure

to fleas as well as to the stool of infected animals. Since most people in the developed world use common, effective, and easily-obtained products (e.g., pills, powders, and sprays) to keep their dog free of fleas, this problem is not widespread in North America and Europe. In any event, the most basic kinds of health practices—cleaning up droppings, keeping dogs free from exposure to feces from unknown, possibly infected dogs, getting rid of fleas, and hand washing are usually enough to prevent problems with these zoonotic diseases.

My Aunt Sylvia shared with others the widespread and false belief that if your family has a pet, your children are more likely to develop allergic responses to animals in general. Doctors commonly recommend that if you come from an allergy-prone family, you should avoid any kind of furry pets. Until recently, some medical practitioners suspected that allergies developed after exposure to the family dog compromised a child's immune system and opened the door to a broad range of general allergic problems—not just those associated with pet hair and dander. However, the truth about dog (and cat) allergies might be the opposite of what experts have long suspected: recent research seems to indicate that raising children alongside furry companions may reduce rather than increase the likelihood that they will develop allergies.

Dennis R. Ownby, the head of the division of allergy and immunology at the Medical College of Georgia in Augusta, whose scientific team published some of this data, described the convoluted path that his group's research took.

"We looked at kids from birth to age seven to see what was the biggest cause of allergies," he said. "We thought high levels of dust mites were probably the number one cause." In addition the researchers looked at a number of other allergy-causing factors, such as parental smoking and pets. The results, however, were not promising.

"Two million dollars and nothing to show for it," Ownby

continued. "Dust mites had no effect that we could find. Then I remembered some other research findings in Southern Germany that seemed to show that kids raised on farms had a lower incidence of allergies. This made me wonder about the effects of exposure to animals, so we went back and looked at the pet questions on the survey.

"The results were really surprising. Our data showed that being exposed to one dog lowered sensitivity to all allergens. Two pets had a bigger effect—producing around 70 percent reduction in sensitivity."

Growing up with pets not only decreases the odds of developing animal-related allergies but also decreases the risk of other allergies, such as dust, and pollen from plants such as ragweed or certain grasses. Children with dogs in their home are also less likely to have eczema, a common allergic skin condition that causes red patches and itching. Because children with allergies are more likely to develop asthma, pets may also lower the odds of having this dangerous respiratory disease. This was partly confirmed by another piece of research that looked at wheezing in infants, which predisposes them to developing asthma later in

Babies who live in a house where there are pet dogs are less susceptible to a broad range of allergies when they grow up.

life. This research showed that infants living with a family dog were one-third less likely to wheeze.

The mechanism by which pets lower the likelihood of allergies later in life probably involves higher levels of substances called *endotoxins*, natural compounds produced by bacteria, in the homes of dog and cat owners. Found in the mouths and saliva of dogs and cats, endotoxins are transferred to humans when your dog licks you or when you come into contact with surfaces that the dog may have licked. Exposure to these substances early in life causes a child's immune system to develop a response system that makes later allergic reactions much less likely. In fact, protecting children from exposure to such substances and cleaning constantly may take away the opportunity for a child to develop these defenses—ultimately making them more vulnerable to allergies when they mature.

Those licks from your dog may prevent other problems as well. Jane Heyworth and her colleagues at the University of Western Australia in Crawley were investigating factors associated with developing gastroenteritis, the condition commonly referred to as "stomach flu." They started with the same kind of beliefs that Aunt Sylvia had, namely, that dogs are dirty and track contaminants into the house and thus may increase the likelihood of this stomach problem. For six weeks, Heyworth's team closely observed 965 children aged four to six and recorded any occurrences of nausea, diarrhea, and vomiting. Instead of finding that pet-owning homes had more symptoms of gastroenteritis, they found that children who lived with a cat or dog in their household were 30 percent less likely to have such gastrointestinal problems. She ultimately concluded that, "Being licked and touched by pets may allow children to develop immunity from repeated, low-level exposure to pathogenic organisms."

Taken as a whole the scientific data say that although having a pet dog in the house may slightly increase the probability of contracting a limited set of zoonotic diseases, living with a pet may actually improve your health on balance. On that summer afternoon those many years ago when my Aunt Sylvia was wagging her accusing finger at my dog, these facts were unknown, so my mother resorted to a different authoritative source to counter her sister's arguments. When Sylvia finally slowed her torrent of criticism and advice and settled down for a cup of coffee, my mother quietly voiced her opinion.

"Sylvie, the Bible has convinced me that our dog is good for my health and for the health of the kids. The dog is a good playmate and makes the boys happy. Truth is, Skipper is a silly little dog who makes me laugh sometimes as well. Remember that the book of Proverbs says that, 'A cheerful heart is good medicine, but a crushed spirit dries up the bones.' So we'll just keep the dog, since God Himself says that there are medical benefits if we do!"

Chapter 27

Physicians and Psychiatrists with Paws

HERE'S A CONSIDERATION that may raise your stress level: You are a hard-driving, ambitious person trying to juggle several projects and meet your deadlines, or you are a homemaker trying to cope with several kids, tight budgets, and the demands of a work-weary husband—and you are going to do this with a body that still thinks you are a caveman! The reason that this is a problem is because we have not had adequate time to evolve physiologically to deal with such problems.

Primitive humans faced stressful situations that were high-threat but of only short duration, such as catching sight of a saber-toothed tiger and dashing for safety. In our modern society we tend to face stressful situations that are of much lower threat (nobody dies because someone missed a deadline for getting work done) but may be of much longer duration (like continuing work stress or a marriage that is on the rocks).

When you are under stress, your caveman body and nervous system still think that you have only two options—fight or run away, both of which require the full involvement of the cardiovascular system. Your heart speeds up to increase the blood flow and carry more oxygen to the muscles and organs so that you can respond more quickly. This increased blood pressure and heart rate are sustained by a variety of hormones released by various glands, including the adrenals. Your body is also preparing for possible injury, so your blood-clotting mechanisms are working at full force, and your total metabolic rate is at high speed. Keep up this kind of physiological activity for hours, and you'll be exhausted. Keep it up for years, and you may hasten a heart attack, especially if you have one or more of the primary risk factors, such as obesity, diabetes, a smoking habit, lack of exercise, chronic hypertension, or a genetic predisposition for coronary artery disease.

The research linking heart problems to psychological stress is impressive. A study recently published in the *International Journal of Epidemiology* for example, involved eight years of testing. It was done in the Whitehall district of London and involved a huge test group (73 percent of all civil servants working in twenty government departments). A variety of stress factors, such as marriage or other family problems, work-related issues, and monetary concerns were considered. The effects of stress were even worse than had been anticipated. Those men who were under psychological stress were 83 percent more likely to have coronary heart disease. Women in the psychologically-stressed group had a still-frightening 51 percent increase in heart problems.

Another larger-scale study conducted in Japan and recently reported in the scientific journal *Circulation* involved more than 73,000 people aged forty to seventy-nine. It found that people who feel stressed on a day-to-day basis have an increased likelihood of dying from stroke or heart disease. Probably the most important finding was that these effects even had an impact on the lowest-risk groups (women without any other risk factors). These stressed-out women were still more than twice as likely to die of heart complications than their more mellow peers over the time period studied.

So what can you do to lower your stress level? Many people resort to psychoactive drugs, including tranquilizers and antidepressants, or alcohol and tobacco, all of which have their own harmful physical effects. Even medically prescribed drugs can lead to dependency and often have negative side effects. Meditation has been shown to be helpful in reducing and dealing with stress. But a much simpler remedy is to get a dog as a pet.

Your dog and, to a lesser extent your cat, can help to tame a stress response that places your health at risk. This medical recognition of the significance of the human-animal bond and its influence on human psychological health is fairly recent. In an earlier era affection between people and their dogs was actually viewed as a sign of psychological illness. For example, on January 29, 1908, an article appeared in the *New York Times* that dealt with the decision of a military tribunal concerning a Colonel Deems and his dog, Riley. According to the article, "The Retiring Board in solemn conclave decided that the Colonel's fondness for the little fox terrier that had the run of Fort Howard, Baltimore, was not an evidence of mental derangement."

The testimony against the officer was supposed to be quite damning, such as, "It must not be forgotten that Riley jumped right up in the Colonel's ample lap and kissed him squarely in the mouth. Did it scores of times. Once, he so far forgot himself as to carry off one of the Colonel's boots surreptitiously. The post commander had to hobble around his quarters for an hour with one foot bootless, while his orderly searched for the No. 10." Further, the article reported the Colonel had done nothing when his dog acted "in utter disregard of the seriousness of army life," by treating officers and enlisted men in exactly the same way. Nonetheless, the army officers and surgeons involved sent Col. Deems back to active duty concluding, "The dog was merely the target for the affection of a lonely army bachelor."

Our view of the human-animal bond has changed quite a bit since that hearing at the beginning of the twentieth century. No one can imagine someone's mental state being called into question in our modern world simply because they showed affection to a dog, or vice versa. Today, our view of the human-animal bond has changed to such a degree that we view dogs and other animals as a means of promoting the mental and physical health of their owners.

The strong connection between humans and animals has been a subject of serious psychological research. Scientific evidence about the health benefits of such a relationship was first published about thirty years ago, when a psychiatrist—Aaron Katcher of the University of Pennsylvania—and a psychologist—Alan Beck of Purdue University—measured what happens physically when a person pets a friendly and familiar dog. They found that the person's blood pressure lowered, his heart rate slowed, breathing became more regular, and muscle tension relaxed—all signs of reduced stress.

A recent study published in the journal *Psychosomatic Medicine* not only confirmed these effects but showed changes in blood chemistry, demonstrating a lower amount of stress-related hormones such as cortisol. These effects seem to be automatic; they do not require any conscious efforts or training on the part of the stressed individual. Perhaps most amazingly, these positive psychological effects are achieved faster—after only five to twenty-four minutes of interacting with a dog—than the result from taking most stress-relieving drugs. Compare this to some of the Prozac-type drugs used to deal with stress and depression, which alter the levels of the neurotransmitter *serotonin* in the

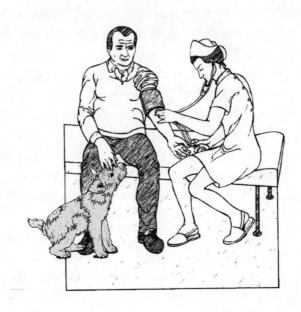

Petting a familiar and friendly dog lowers a person's blood pressure.

body and can take weeks to show any positive effects. Furthermore, the benefits that build up over this long course of medication can be lost with only a few missed doses of the drug. Petting a dog has a virtually immediate effect and can be done at any time.

A large body of data now confirms that pets are good for the health of your heart and may increase the quality of your life and your longevity. The benefits are not just short-term—they reduce your stress beyond the period of time that your pet is present; and seem to have a cumulative effect. For example, one study of 5,741 people in Melbourne, Australia, found that pet owners had lower blood pressure and reduced cholesterol levels than nonpet owners, even when both groups had the same poor lifestyles, involving smoking and high-fat diets.

A fascinating study, presented at an American Heart Association scientific conference, demonstrated how the addition of a pet to your life can help. Researchers used a group of male and female stockbrokers, who were already beginning to show the effects of stress and were candidates for medication to lower their blood pressure. The researchers first evaluated the brokers' blood pressure under stressful conditions by giving the research participants speeded numerical tasks and asking them to role-play a situation in which they had to talk their way out of an awkward position. In response to these stressful tasks, their average blood pressure readings shot up to 184/129 mm of mercury (any blood pressure of 140/90 mm of mercury or more is considered high).

Each of the stockbrokers then was prescribed the same medication, and half of them also agreed to get a dog or a cat for a pet. Six months later the researchers called the test subjects back and gave them additional stress tests. Those stockbrokers who had acquired a pet were allowed to keep the pet with them when they took their stress tests. These pet owners now showed a rise in blood pressure that was only half that of the brokers who were treated only with the medication.

Pets can actually help even if you have started to show evidence of heart problems. In an intriguing study published in the *American Journal of Cardiology* researchers followed more than 400 patients when they were released from the hospital after having a heart attack. One year later the pet owners had a significantly higher survival rate than nonpet owners.

Even though stress is a major contributor to cardiovascular problems, it seldom appears as a symptom by itself. Stressed individuals often feel that their lives are out of control, which leads to a variety of anxiety-related psychological states, most of which commonly include depression. Up to 25 percent of people who seek the services of a general practitioner do so for depressive and anxiety disorders. Depression is considered to be much more disabling—socially and physically—than many chronic physical illnesses, such as diabetes, arthritis, and back pain.

Although depression can be caused by many factors, one of the most important is loneliness due to social isolation. People who lack human contact often benefit from pet ownership and the emotional bond pets provide. Older people are particularly at risk of becoming lonely, isolated, and depressed. Recently, researchers looked at a group of people aged sixty and older, living alone, but half of the group surveyed had a pet. Non pet owners were four times more likely to be diagnosed as clinically depressed than pet owners of the same age. The evidence also showed that pet owners required fewer medical services and were more satisfied with their lives.

The relaxed relationship most people have with their pets seems to be particularly beneficial for people living alone. It increases the likelihood of meeting people since strangers are much more likely to stop to talk with pet owners—mostly because there is a dog to say hello to. People seem to enjoy a moment of relaxed interaction with a pet. This can sometimes have important implications for the person's future lifestyle. Take Emma Cooper's case, for instance. Emma, who is seventy-one, had been living alone in Chicago for nearly eight years after her husband's death.

"I was out walking Surrey—my cocker spaniel—and this man stopped to give him a pat. He seemed like a nice man and told me he used to have a blond cocker spaniel just like Surrey. We started to talk about living with dogs and then stopped for a cup of coffee. Well, one thing led to another and Bill and I are getting married next month—as soon as we can find a clergyman who is willing to let a dog stand in as the best man!"

The reasons behind the medical and psychological benefits of interacting with dogs are still being explored, but it seems that the uncritical and always available social support provided by a pet is one major factor. Psychological research has shown that people who have a lot of social support from family and friends are more stress resistant and suffer from fewer problems that can be traced back to stress and depression. Dogs are always willing to provide social interaction and support and thus provide the same benefits that having caring humans in your life does. This also probably explains why dogs have more beneficial health effects than cats. It is simply because dogs are more social, attentive, and interactive. Nonetheless, cats also produce good outcomes, pet birds do some good, and you can even get some stress reduction watching an aquarium full of fish.

Whatever the reason, when your life becomes sufficiently stressful to have health consequences, you may one day hear some enlightened doctor tell you to "Pat two puppies and call me in the morning."

Chapter 28

Best Friends and Bed Partners

JUST WHOM ARE you sleeping with right now? No, this is not a *Cosmopolitan* or *Playboy* survey about your sex life, but rather a question of whether your four-legged Lassie (or Rover) is snuggled in bed next to you. One recent survey found that about half of all dog owners allow their dogs to sleep on the bed with them. Lassie's chance of getting under your blankets depends upon your age and sex. The highest percentage of people sleeping with their dogs are single females between the ages of 18 and 34. Nearly six out of ten women in this group allow the dog on the bed. The group with the largest likelihood of booting the dog out of the bed are married men over forty-five years of age. However, even for this class of people, just shy of 40 percent still sleep with their dog.

Many famous, rich, and powerful people have given their dogs bed privileges. According to information carved into his tomb, the Egyptian pharaoh, Ramses the Great, had a hound named Pahates whom he gave the title of "Bed Companion to the Pharaoh." Alexander the Great was known to have rested from his battles by sleeping beside his longhaired Grecian greyhound, Peritas. Much later Queen Victoria actually died in bed next to her favorite Pomeranian, Turi. Peter the Great, czar of Russia and Frederick the Great of Prussia slept with their Italian greyhounds (Lissette and Biche respectively), and so it went with many great people.

For most people, having a dog in bed is psychologically comforting. There is a loving companion close at hand, and it keeps you from feeling lonely or insecure no matter how dark the night. From the dog's point of view, however, there are potentially some unwanted implications that result from this sleeping arrangement.

Psychologists have learned that for dogs there is an almost rigid and linear pack hierarchy. It mimics the military in that each individual has a rank, with the pack leader being the general, and so on down to the lowest-ranked individual. The leader (or *alpha*) dog demonstrates his status by controlling resources and occupying positions of power. Alpha goes where he wants and chooses where he wants to sleep. Usually this will be the highest elevation, which gives him a chance to look down on the others (as if he were a very big dog). Therefore, if you sleep on the bed and the dog sleeps on the floor, you subtly assert your position as pack leader. If you allow the dog onto the bed, however, he has the same elevation as you do, which implies the same power position and equal status. This can undermine your leadership role with the dog and may result in demanding behavior or outright disobedience on the part of the dog.

There is an additional pitfall. Pack members acknowledge the leader by moving out of his way when he approaches. Any animal that does not move is physically bumped out of the way. By being forced back a step or two, a dog in effect admits that he is lower in status. A subtle variation of this kind of behavior is leaning, which is really nothing more than a very passive version of the shoulder bump. A dog that wishes to express its dominance will lean its weight on another dog. If that dog moves however slightly, it has conceded the higher status to the dog doing the leaning. Unfortunately, when we are in bed with our dogs, if they lean on us we may out of politeness and a wish not to disturb them move away to give them more room. Again this is a signal that is read by the dog as evidence that they are really higher in status than the person who has moved. Thus if you *must* sleep with your dog, at least force the dog to move when you wish to turn or get more space. This will help to undermine any delusions of grandeur that Rover might be tempted to develop.

There is another potential problem with having Lassie as a bed partner. The survey also found that 13 percent of the couples studied included a partner who so objected to the dog being on the bed that it had actually become a point of controversy and emotional strain in their relationship. A famous case is that of the Indian fighter, General George Armstrong Custer. Custer had frequent heated disputes with his wife Libbie over the presence of dogs on their bed. It eventually came to the point that she threatened to no longer sleep beside him if it involved sharing the bed with his dogs. They eventually reached a compromise agreement. When Custer was at home the dogs could sleep in their bedroom but *not* on their bed. In the field, however, Custer shared his cot with his greyhounds Blucher and Byron, and the fawn-colored deerhound Tuck, the dog that died with him at the Battle of Little Big Horn.

There appears to be one further potential pitfall to sleeping with your canine friend. Some dogs are restless sleepers and move around a lot, shifting positions many times during the night. This kind of activity can disrupt your sleep quite a bit. Yet another survey of the effects of dogs-on-the-bed found that people reported that, on average, they were awakened at least once each night due to movement by their dog, or because they had bumped up against the dog or because the dog was crowding them in their sleep. While a single wakening is not much of a sleep disturbance, some dogs are considerably more restless. In the group of people surveyed, close to 20 percent said that they were usually awakened by the dog three or four times per night, and about 4 percent of the respondents admitted that their dog could sometimes awaken them eight to twelve times in a single night. This is enough sleep disruption to have significant consequences the next day. Individuals with this degree of sleep loss might be expected to have slower reflex responses and some impairment in their short-term memory and problem-solving ability. Under certain circumstances this might have dire consequences.

Take as an example the case of Baron Manfred Freiherr von Richthofen, who came to be known to the world as the "Red Baron" and would go down in history as Germany's greatest World War I fighter pilot. Richthofen began his service as a fighter pilot in September 1916 and shortly thereafter became commander of Fighter Group 1. He was proud and flamboyant, and so that there would be no mistaking him in combat, he painted his Fokker triplane a bright red. It was this that gave him the nickname of *der rote Freiherr*, or "Red Baron." Following the example set by their commander, his entire fighter group painted their planes in bright colors, resulting in his squadron's nickname, "Richthofen's Flying Circus." The visibility of his red plane at a great distance did not affect his success in combat. Richthofen was a deadly fighter, and one of the top flying aces of all time. He was personally credited with shooting down eighty enemy aircraft.

During his years as a pilot, Richthofen shared his life and his quarters with a Great Dane named Moritz. At all hours of the day, the tall aviator could be seen wandering around the airfield in his leather jacket, with his hands in his pockets and his large dog marching by his side. On several patrol and observation missions, Richthofen actually strapped Moritz into the second seat of his plane and took him along (which the dog apparently did not like at all).

Moritz, however, was extremely loyal to Richthofen. He would go out to the field with him before each mission and sit and watch him take off for battle. He would then wait patiently near the runway for his return. There was no radar at that time, and the only way to know that the planes were returning was to visually sight them. However, Moritz's sensitive ears would detect the approach of his master's plane well before it could be seen. He would signal this by moving toward the runway, barking as he went. At his cue the plane spotters would immediately start scanning the horizon in the direction that the big dog was looking, since he was almost always correct.

Baron Manfred von Richthofen (the "Red Baron") and his Great Dane, Moritz.

Richthofen often took Moritz with him to various social occasions or simply when he was spending time with other pilots. It was not unusual for him to share a beer with the dog at the officers' club. It is even said that on the same shelf where the pilots kept their personal, ornate beer mugs, there was also a shallow blue bowl with vertical sides and a gold "M" emblazoned on it, and this was for Moritz's portion of beer. The big dog resided in Richthofen's quarters with him and actually slept on his bed. It may well have been this fact that was Richthofen's undoing.

The Battle of the Somme was in progress, and on the morning of April 21, 1918, Richthofen went out to the airstrip to board his plane. He was yawning and looked fatigued. One of his squadron commanders, Lieutenant Hans Klein asked him, "You look tired. Were you out late last night?"

Richthofen gave a small laugh and replied, "Moritz was very restless last night. He kept waking me, and I doubt that I got two hours' worth of real sleep. They will have to give me a bigger bed if they expect me to get enough rest to win this war."

Later that day the famous red triplane sustained damage from ground fire. In a now partially crippled plane, Richthofen had the unfortunate luck to encounter a Sopwith Camel flown by one of Canada's top air aces, Captain A. Roy Brown. This would be the one dogfight the Red Baron would lose.

It is intriguing to wonder about Richthofen's final night with Moritz. Could it be that the dog's restlessness caused the great pilot to lose enough sleep so that on this day the Red Baron's reflexes and judgment were slightly impaired? We will never know, though it is an interesting speculation.

In any event the cases of General Custer and Baron von Richthofen should serve as a warning to tread lightly in this matter. Your selection of bed partners can have a powerful influence on your life—even if that bed partner has four legs and a tail.

Chapter 29

Can Dogs Help Fight Cancer?

BRUCE BLACK IN Toronto still walks on two legs, while Arlene Goldberg in Chicago has been cured of her potentially-fatal cancer—both because of dogs. The two cases are quite different, however. Bruce's recovery was possible because of medical work done *with* dogs. Arlene's recovery was possible because of work done *by* dogs.

Let's start with Bruce. Traditionally, research on cancer has used rats and mice. Unfortunately, scientists have found that many drugs that help cure the artificially-induced cancer in rodents don't work as well with humans. For this reason, in 2003, the National Cancer Institute began the Comparative Oncology Program, which designs trials involving dogs with naturally-occurring cancers.

There are a number of reasons why this made sense. First, cancer is quite common in dogs. It is the leading cause of disease-related death in dogs older than age two and 50 percent of dogs older than ten will be diagnosed with cancer. Second, dogs and people develop the same kinds of cancer, with similar disease patterns. Dogs and humans are both plagued by bladder cancer, melanoma, lymphoma, and mouth and sinus cancers. They are the only two species that naturally develop lethal prostate cancers. Another parallel is that the most frequent bone cancer in pet dogs is osteosarcoma, the same cancer that strikes human teenagers (like Bruce). A look through a microscope shows that cancer cells from a teenager with osteosarcoma are identical to bone cancer cells from a golden retriever.

Dogs are a good choice to use as a model for human cancers at the genetic and evolutionary levels as well. Dogs are closer genetically to people than mice. Because the canine genetic code has already been sequenced, we know that the genetic codes for dogs and people overlap a great deal. This works for not only major physical factors but for subtle ones—thus the gene that causes brown eyes in people is the same gene that causes brown eyes in a dog.

Purebred dogs have an extra advantage. Humans, much like "mutt" dogs, breed randomly, while pedigreed dogs have controlled breeding so that their DNA remains relatively unchanged. That genetic purity cuts down on background noise within the genome, making cancer-causing mutations easier to identify. In humans, searching for cancer-causing genes is like trying to listen to a radio that's out of tune with lots of static. In pedigreed dogs a lot of that background interference is removed, allowing scientists to pinpoint a particular cancer-related gene—much like tuning in a radio.

The evolutionary model that humans are based on is also similar to that of dogs. Biologists suggest that evolution designed us to live long enough to successfully reproduce and rear our children. Unfortunately, after that stage is over, evolution has no concerns about how rapidly we fall apart. Such a design makes us poorly equipped to resist or repair the damage that accumulates in our bodies. Over time the deterioration can eventually unbalance cells enough to result in cancer. In the distant past, our ancestors, and those of our dogs, did not normally live long enough to become afflicted with such age-related cancers.

Dogs share something else with humans that makes them ideal models for cancer research: the environmental risk factors for human cancers are virtually duplicated in a dog. Unlike lab rats or mice, pet dogs live in our homes, breathe the same indoor air we do, drink the same water, play on the same grass, and are exposed to the same living conditions as their human owners. Lab rats don't *get* cancer, they are artificially *given* cancer. Because dog studies are based on data about pet dogs living in their families' homes, the cancers are naturally occurring. The result is these pets act like "environmental watchdogs" that can help identify cancer-producing substances in our lives. Most cases of mesothelioma, for example, which is a cancer affecting the tissues lining the chest and abdomen, result from contact with asbestos. Symptoms are slow in appearing and may take up to thirty years after exposure to appear. Mesothelioma in dogs is also largely related to encounters with asbestos, probably through being near their masters' work or hobby use of the product. In dogs, however, the time between exposure and diagnosis is much shorter—less than eight years. This means that the appearance of mesothelioma in a dog can alert people to look for and remove any remaining sources of asbestos. Furthermore, careful monitoring of the dog's owners can lead to earlier diagnosis of the cancer, while it is still curable.

In North America people are increasingly willing to pay for treatments that prolong their pets' lives. In the United States alone, dog owners spend well over $20 billion annually for veterinary services, and much of this is for cancer treatment. Pet owners are often quite willing to enroll their dogs in clinical trials and studies if these may ultimately benefit their own dog. Generally, owners are required to pay for diagnosis and the tests necessary to see if their dog is eligible to participate in the research program. After that, all costs are typically covered by the research organization or drug company. The added bonus is the possibility that the new treatment may help their pet have a longer, more comfortable life. In any event, even when the chance of helping their own dog is remote, if the research might help other dogs as well as human beings in the future, people seem to be sufficiently motivated to enlist their dogs in such scientific projects.

Notable progress has already been made. Between 1975 and 1995 the incidence of bladder cancer in dogs in North America increased sixfold. Although the data suggested that terriers might be more susceptible to the disease, there was no apparent reason for the increasing rates. However, when researchers interviewed the owners of Scottish terriers with bladder cancer, they found that dogs whose owners had used phenoxy acid herbicides on their lawns were more than five times more likely to have cancer than dogs whose owners had not. They also found that diet could have a protective effect, since Scotties that ate green, leafy vegetables three times a week showed a 90 percent reduction in their risk of cancer. The human implications are obvious.

Researchers at the Animal Medical Center and Memorial Sloan-Kettering Cancer Center in New York have developed a vaccine that works for melanoma. It uses DNA from mice or humans to trigger an anticancer-immune response in dogs. The key ingredient is tyrosinase, a black pigment found in melanoma cells. Injected into a dog, the tyrosinase changes into a protein that the body sees as foreign because it is from a different species. As the immune system responds, it attacks the tyrosinase found in the melanoma, breaking it down. In four clinical tests the drug quadrupled the average survival time of dogs with the disease, eradicated some cancers, and had few side effects. It is already licensed for animal use and, based on the canine data, the U.S. Food and Drug Administration has granted permission to begin human trials.

All of this brings us back to Bruce and the osteosarcoma discovered on his leg. Twenty-five years ago a diagnosis of osteosarcoma in a teenager meant amputation of the leg, and without effective drugs to attack the tumors, almost certain death. Research done on golden retrievers by Stephen Withrow and his colleagues at Colorado State University changed this. First, the researchers showed that chemotherapy delivered directly into an artery could convert an inoperable tumor into an operable one. Amputation was avoided by chipping out the diseased bone tissue and replacing it with a bone graft and metal implant. So, because of those golden retrievers Bruce can still walk on his own two legs.

Arlene's case was quite different. She owes her life directly to her cocker spaniel, Duffy, a dog that developed an annoying habit of jumping up on her and sniffing and nipping at a mole on the back of her shoulder near the base of her neck. He was so persistent that Arlene mentioned the mole to her doctor. He removed it and had a biopsy done. A few days later, Arlene was back in the hospital to have the surrounding skin removed because the mole turned out to be a virulent form of melanoma which could have killed her had it metastasized.

For many years there have been anecdotal reports of dogs that detected cancer in their owners, and recent, controlled studies confirm that dogs can detect cancer as well as or better than traditional medical-screening techniques.

Carolyn Willis of Amersham Hospital in England, and her associates, for example, conducted a meticulously-controlled, double-blind study showing that dogs can be trained to recognize and indicate bladder cancer. They used sets of six urine

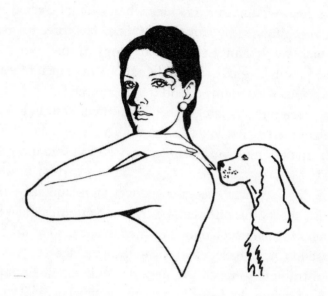

Duffy kept sniffing and nipping at a mole on the back of her shoulder.

samples belonging to patients who either were healthy or suffered from some other disease, plus a sample from a patient with bladder cancer. Neither the researchers doing the testing nor the six dogs had any way of knowing in advance which sample was cancerous until after the dogs made their choices.

In one instance the dogs kept identifying a sample that medical staff had asserted was cancer free. Willis said, "The trainers just couldn't train the dogs past this particular sample at all. They were really getting quite desperate that this wasn't going to work. Because the dogs consistently went for that one sample, we went back and conferred with a specialist."

Andy Cook, one of the trainers, describes what happened next. "The hospital had seen our dogs' work and had got confidence in our dogs, so they sent it off for further tests. They were completely blown away when it came back that this patient not only had cancer on his kidney but bladder cancer."

Willis reported another instance where a dog detected cancer that had been missed by a doctor. "One of the three breast cancers, which we've had picked up by dogs, turned out to be a very, very small focus of malignancy, undetectable unless screened. This was removed, and the dog immediately lost interest. But three months later, it began sniffing, snuffling and becoming agitated again when sitting on her lap. So she shot back to the hospital, and lo and behold, they had missed a tiny bit of cancer!"

Some dogs seem to recognize cancer spontaneously, but recent work shows that they can be trained to do this quickly and effectively. Michael McCulloch of the Pine Street Foundation in San Anselmo, California, and his colleagues needed only three weeks of training to teach five pet dogs to detect lung or breast cancer by sniffing the breath of participants. The trial itself involved eighty-six cancer patients (fifty-five with lung cancer and thirty-one with breast cancer) and a group of eighty-three healthy patients. In the study, the dogs sniffed breath samples captured in special tubes. Dogs were trained to sit or lie down directly in front of a test station with the cancerous sample. The results were spectacular, showing that dogs can detect breast and lung cancers with an average of better than 90 percent accuracy.

Perhaps sometime in the future that lab test that you get for possible cancer may well come in the form of a black Labrador retriever!

Chapter 30

Guardian Angel

A WOMAN WE'LL call Nancy was wading along the shore of Lighthouse Park, looking for a purple starfish that someone had claimed to have seen in the area, when a white bag fell from the sky, apparently thrown from the roadway that skirted the high cliff behind her. She watched it sail downward and was startled by a yelp or cry of pain that it made when it hit the surface of the water.

Nancy is a biological researcher at the University of British Columbia with the usual detached and relatively unemotional personality style that many scientists—especially those who work with physiology and the machinery of life—often develop. However, that does not mean that she lacked empathy or caring. So she quickly waded out the few meters to where the bag had sunk from sight. Reaching down, she pulled up what appeared to be an old pillow case or laundry bag that had been tied closed with some string. She could hear choking and gagging sounds from whatever creature had nearly drowned inside of it. Nancy's cautious scientific mind-set was now completely abandoned. There was a small rip in the cloth and she worked her fingers into it and pulled as hard as she could. The cloth tore open and she could see wet white fur. She ripped a wider opening, and out of it emerged a white face with pricked ears, coal-black eyes, and the equally-dark nose typical of a West Highland white terrier.

She clutched the small wet white dog to her chest and stumbled from the cold water to the shore. She could feel his warm tongue licking at the side of her face, and when she finally put him down on a smooth rock surface he would not stop licking her hands in what appeared to be sincere gratitude. When she examined him, she concluded that he was a relatively unharmed dog, perhaps a year or less in age, a bit small for his breed, but with all of the personality characteristics that define a terrier. He wore an inexpensive cloth collar which had a tag hanging from it that read "Angel."

Nancy lived alone in a two-story townhouse. The long hours that scientists often put into their work frequently make it difficult to establish a wide circle of friends or close relationships. Furthermore, she had found that many men are wary of or insecure around bright, well-educated women with advanced degrees. Living without a companion had been wearing on her, and Nancy had been thinking about getting a pet. Now here was Angel. His arrival seemed like providence. She looked at the little dog that was frantically licking her fingers and wagging his carrot-shaped tail and announced, "You need a home? You've got one!"

I am sure that if one could read the mind of a dog, one

would have heard Angel reply, "You need a guardian? You've got one!"

Angel immediately set himself up as Nancy's protector. Anything that encroached on her territory required that the alarm be sounded and a full-out attack launched by her little white knight. The first time that a neighbor's cat entered her small front yard, for example, Angel began to bark frantically and charged the orange-and-white intruder. The cat leapt back over the seven-foot-high fence, but Angel was moving too fast to stop, and in defiance of gravity made it to the top, then pitifully hung there by his front paws. Nancy ran over to grab him before gravity could take revenge and slam him to the ground. As she put him down, she giggled, "My hero!"

Friends or family arriving to visit had their way noisily barred by Angel until Nancy arrived to usher them inside. Once Nancy acknowledged and accepted visitors, they would be treated with the utmost courtesy by her furry sentinel and would be recognized as welcome guests on future visits.

It might be said that Angel was hypervigilant in his quest to guard the woman who had saved his life. Anything new or strange was treated as an intruder. One day, Nancy bought a pair of lawn chairs and a little table so that she could sit outside in her little yard and read in the summer sunshine. The moment that Angel encountered the new pieces of furniture, he decided that they were alien monsters threatening his mistress. He braced his legs and barked and growled a mighty threat. He interposed himself between Nancy and the green wooden invaders and would not let her approach this new danger. Ultimately, she had to pick him up and sit down on the chair with him on her lap to convince him that she had the situation under control.

The truth is that Nancy had always felt the typical insecurities of women living alone in the city. However, she now felt a lot safer with Angel around to sound the alarm. He was particularly vigilant at night, when he slept in her bedroom. A tree limb brushing against the house might be enough to cause him to waken his mistress with warning barks. Although there were many false alarms, twice Angel's nighttime alerts proved to be extremely valuable. Once, when Angel woke her and she went to the window, she saw someone trying to get into her car parked on the street. A shout from her caused the potential car thief to flee. Another time, during a rash of break-ins involving condominiums and townhouses that had ground-level access in her area, she was again aroused by Angel's noisy alarm-barking. Looking out the window, she saw two men in her yard. A quick 911 call to the police resulted in their arrest, and they were later charged with a number of local burglaries and home invasions. Angel got a cookie and a warm hug for that action.

A few times, Angel's protection was too zealous. One day a jogger trotted by Nancy's yard, and Angel barked a defensive challenge and rushed at the garden gate. It had an old-style latch consisting of a short metal bar that rested in a notched flange. When Angel hit the gate, the bar bounced and slipped out of the notch, allowing the white dog to shoot out of the yard and down the street. He launched himself into the air and bit into the jogger's light cotton running shorts, ripping out the entire rear portion. Nancy raced through the gate and grabbed her dog. She apologetically ushered the jogger into her home. Fortunately, Angel had only bitten into the cloth, not flesh, and while the jogger sat on her sofa with a towel to cover himself, Nancy hastily stitched a repair. Angel now seemed reconciled with the new "visitor" and worked hard enough at being friendly so that he eventually solicited some friendly pats from the runner.

It turned out that the jogger's name was Robert, and he was a paramedic. He and Nancy had a pleasant talk, and after she drove him home, they exchanged phone numbers. Seven months later, Angel was led down the aisle by Nancy's nephew as the ring bearer at Robert and Nancy's wedding.

There were times when Angel's protection was too zealous.

Robert was not fond of Angel's late-night alarms—which he called the "game of imaginary burglar." This was mostly because the feisty dog would start his warning barking out on the balcony, and on damp nights when he jumped onto the bed to alert his mistress, he would leave muddy footprints on the bedclothes. Nancy took the problem in stride; she simply put a top sheet over the blanket and washed it whenever her defender's pawprints made appearance.

All of Robert's objections to Angel's behavior disappeared when he proved how he truly was Nancy's guardian. One evening she had gone visiting and had taken Angel along to play with her sister's Labrador retriever. It was late at night when she parked her car, put a leash on Angel, and started across the street to her home. Suddenly a man stepped out of the shadows, waved a knife, and shouted at her. Angel shot forward and bit the man in his leg. The pandemonium that resulted gave Nancy enough time to grab a canister of pepper spray from her purse and to spray his face. As he roared in pain, she and Angel dashed to their door and made it safely inside. Angel was bleeding from a shallow cut, but his mistress was safe. She called him her "guardian angel," and Robert admitted that, all things considered, he could live with those muddy footprints.

Time passed as it always does, but age did not daunt the spirit of Nancy's defender. Then one day the barking stopped, and Nancy buried a small urn filled with the ashes of her little white shield and companion in the corner of their garden. Robert stood by trying to comfort her while she repeated over and over, "He was always there to protect me." Robert silently grieved for his wife's loss and made plans to get Nancy another dog after a respectable amount of time had passed.

It could not have been more than a month later when Nancy awakened to the sound of loud barking. It sounded just like Angel's typical warning. She rolled over in bed to ask Robert about it, but then remembered that he was on night call that evening. The barking was so intense that she sat up to try to figure out where the sound was coming from, but when she did, she felt a wave of nausea and found herself gasping for breath. In her mentally fogged state, she could only think that Angel was worried about something and trying to alert her.

Nancy pulled herself to her feet. It was hard to breathe, and she was disoriented and dizzy. Yet there was that barking again, now coming from the hall. She staggered from the bedroom following the frantic sounds of the dog, virtually tumbled down the steps, and found herself at the front door. Now the barking was coming from just outside, so she fumbled with the latch to get to the dog. She felt weak. She could hear herself gasping for breath, yet somehow she opened the door, took two steps forward, and collapsed in a heap.

A neighbor, Andrew, heard the commotion and looked out. He saw Nancy lying on the grass and rushed out while his wife called for help. When the police and emergency crew arrived, they found that there had been a major gas leak that had filled the house with deadly fumes.

As fate would have it, the paramedical van that came to her assistance was Robert's. After the administration of some oxygen, Nancy was recovering quickly and Robert could only mumble about how lucky she had been to make it out safely, given how she had been so soundly asleep. "So many people just don't wake up when this sort of thing happens," he said.

Nancy leaned against him and said "It was Angel who woke me. I know that sounds stupid, but I know his bark. He led me to the door. Andrew heard the barking, and that's what woke him as well."

Robert looked at the neighbor who stood nearby and was simply nodding agreement. Robert looked for a more rational explanation, and he suggested, "It was probably just another dog in the area, and just a lucky thing that it was loud enough to wake you up."

By now the gas had been turned off, and the house was finally clear enough to return. Robert walked Nancy upstairs encouraging her to try to get some sleep and assuring her that he would be back home in an hour or so. They entered the bedroom and turned on the lights. It was then that they saw the muddy pawprints tracked across the bed.

Chapter 31

Are There Dogs in Heaven?

MARTIN LUTHER, THE German priest and scholar whose questioning of certain Church practices led to the Protestant Reformation, had a daughter named Mary Catherine and a dog named Tolpel. One day Mary came into his study with Tolpel. Mary loved the dog very much, but he was growing old and frail.

"Father," she asked, "What happens when my dog dies? Does he go to heaven?"

This question has been asked many times, not just by children, but by adults, scholars, and clergymen. In early history there was no question that dogs had souls and would be allowed into heaven. Ramses III, who became pharaoh of Egypt in 1198 B.C., interred his favorite dog, Kami, with all the ritual ceremony due to a great man, including a coffin, linen, incense, and jars of ointment, plus the ritual scroll he would need for his entry into paradise.

In later religions, dogs actually became *psychopomps*. This meant that when a person died, it was a dog's job to escort him to the next world and protect him and show the way. Yima, the Zoroastrian god, was believed to have set two four-eyed dogs to guard Chinvat Bridge, known as the "Bridge of Decision," between this world and heaven. These dogs were placed there because like all dogs, they were good judges of character and would not let anyone pass on to paradise if they had deliberately harmed a dog in this world.

The rise of Christianity seems to have ushered in the belief that dogs would not make it to heaven. Despite the fact that the word *animal* is derived from the Latin word *anima,* which means "soul," Christianity has traditionally taught that dogs and other animals have no more consciousness or intelligence than rocks or trees. According to religious doctrines at the time, anything that had consciousness also had a soul, and anything that had a soul could earn admission to heaven. This was simply unacceptable to the Catholic Church. In later years they would claim support from certain scientists and philosophers, such as René Descartes, who would have described your dog as just some kind of machine, filled with the biological equivalent of gears and pulleys. This machine doesn't think but can be programmed to do certain things. Machines have no souls, and therefore one need not allow a beagle-shaped automaton or a mechanized Maltese to pass through the pearly gates of Heaven.

These views were strongly held, and Pope Pius IX, who headed the Church longer than any other pope (1846–1878), actually led a heated campaign to try to prevent the founding of the Italian Society for the Prevention of Cruelty to Animals, on the grounds that animals have no souls. Pius quoted Thomas Aquinas to prove his case, since Aquinas often noted that animals are not beings, but just "things." Aquinas, however, seems to have had some doubts when he warned that "We must use animals in accordance with the Divine Purpose, lest at the Day of Judgment they give evidence against us before the throne." Such words certainly suggest that animals would be around in the afterlife.

It is interesting to note that Pope Pius (who created the doctrine of Papal Infallibility) was contradicted in 1990 by Pope John Paul II, who said that "also the animals possess a soul, and men must love and feel solidarity with smaller brethren." He went on to say that animals are "as near to God as men are." Although the Pope's statement was reported in the Italian press, it appears that it was not widely discussed—perhaps to prevent the embarrassment of having two infallible popes contradicting one another, or to avoid a class action suit that might be filed against the Church by the spirits of animals that had been unjustly denied entry into heaven in earlier regimes.

One should not blame the Church, however, considering that the Bible is silent on the matter of whether our dogs will make it into heaven. There are some hints, however, particularly in the Apocrypha, which is a collection of books that don't make it into all Bibles and are recognized by some sects but not by others. One is the Book of Tobit, which describes how Tobias went off on a trek to collect a debt to help his blind father. He was accompanied on this journey by the angel Raphael and a small dog. After all their adventures were finished, Tobias returned home, and the dog ran ahead to announce his arrival. Tradition maintains that this dog even preceded Tobias into heaven. Actually, it is this story that accounts for the sustained popularity of the name Toby for dogs.

The actions of individual saints also suggest that some believed that dogs would be in heaven. According to Irish folklore, Saint Patrick repaid the legendary character Oissain, who helped him set up the Church in Ireland, by assuring him that he could

It is said that Tobias's dog preceded him to heaven.

take his hounds to heaven with him when he died. I suppose that they are still romping with Tobias's little dog there.

Muslim dogs also face ambiguity when it comes to dogs going to heaven. The Hadith, which is a collection of traditions associated with the Prophet Muhammad, tells us that angels won't enter a room if a dog is there, and anything a dog touches is tainted and must be washed seven times. However, in the Qur'an itself, there is the *Story of the Seven Sleepers*. The incident began during the reign of the Roman Emperor Decius around A.D. 250. In an effort to strengthen the state-supported religion, non-believers were systematically persecuted. Seven faithful young men fled to a cave on Mount Coelius accompanied by a pet dog. Once in the cave, some feared that the dog—Kitmir by name— might bark and reveal their hiding place, so they tried to drive it away. At this point, God granted the dog the gift of speech, and he said, "I love those who are dear unto God. Go to sleep, therefore, and I will guard you." After the men had settled down to sleep, leaning on the back wall of the cave, the dog stretched out with his forelegs facing the entrance and began his watch.

When Decius learned that there were religious refugees hiding in local caves, he ordered that all entrances be sealed with stone. Kitmir maintained his vigil, even while the cave was being sealed, and made sure that no one disturbed the sleepers. The men were forgotten, and they slept for 309 years. When they were finally awakened by workers excavating a section of the mountain, the dog finally stirred and allowed his charges to return to the world, since it was now safe for their faith. According to Muslim tradition, upon Kitmir's death, the dog was admitted to paradise.

I encountered my favorite analysis of the question of whether there are dogs in heaven when I was in the army and stationed at Fort Knox, in Kentucky. I spent a good deal of my free time in the surrounding countryside, talking with people about dogs, and on one such day I ran into a man who had some interesting-looking hounds. As he described the history and breeding of his unique animals, he motioned me over to some heavy wooden rocking chairs on the porch and offered to get me a beer. It turned out that he was a Baptist minister in charge of the little church next door, and he had the appropriately biblical name of Solomon, although I have no recollection of his last name. As we sat and talked, he commented, "Yep, these will be the most handsome dogs in all of heaven."

I commented, "So as a man of God, you feel that there will be dogs in heaven?"

Solomon smiled a slow smile, and started to speak in that singsong cadence that the clergy always use in their sermons.

"Let me tell you, brother, it is a real arrogance that we have that says that only humans have a soul—that only humans can go to heaven. Are we special simply because we rear up and walk on our hind feet? Is our mouth any closer to the Lord's ear than the dog's mouth, simply because his front paws are still planted on the ground and ours are in the air? Are we worthy of some special salvation, or to have a divine afterlife reserved for us alone, just because we wag our tongues instead of our tails? I think not. If a dog is good and keeps the faith the way a dog should, by doing what a dog is supposed to do, is there any reason why he should not be in heaven? And even more than that, would good God Almighty try to convince us that an existence without the company of dogs could really be heaven? No sir—if there are no dogs in heaven then I don't want to be there. I tell you, dogs are a blessing, and since heaven is for the blessed, there certainly must be a whole lot more dogs than people on the inside of those pearly gates."

Solomon's arguments did not depend much upon scripture, or the formal dogma of the Church. He spoke mostly as a man with a certain faith and a belief that a just God would grant a good person the company of dogs. It is a view held by many people, such as Robert Louis Stevenson, author of novels such as *Treasure Island*, who declared, "You think dogs will not be in heaven? I tell you, they will be there before any of us."

My own view is much like Solomon's. For those who love dogs, it would be the worst form of a lie to call any place where dogs are banned "Paradise." Certainly no loving God would separate people from their canine friends for eternity. If there are no dogs in heaven, then for me there isn't any heaven.

All of this brings us back to Martin Luther and his daughter, Mary Catherine's, question about whether her old dog, Tolpel, would go to heaven when he died. As his answer, Luther rose from his desk, walked over to the dog, and bent down to pat it. As he did so, he spoke to Tolpel with great assurance, "Be comforted, little dog. Come the resurrection, even thee shall wear a golden tail."

Endnotes

THIS IS A LISTING of some of the sources for material discussed in this book. I have tried to include review material whenever possible so that readers can use these citations as a starting point for finding the more specific research they may be interested in. If a reprint or a later edition of a work was consulted, that source, rather than the original, is listed. It also includes a listing of the original *Modern Dog* magazine articles that were the starting point for several of the chapters in this book.

Bekoff, M., C. Allen, & G. M. Burghardt. *The Cognitive Animal: Empirical and Theoretical Perspectives on Animal Cognition.* Cambridge, MA: MIT Press, 2002.

Brenoe, U. T., A. G., Larsgard, K. P. Johannessen, & S. H. Uldal. "Estimates of Genetic Parameters for Hunting Performance Traits in Three Breeds of

Gun Hunting Dogs in Norway." *Applied Animal Behavior Science*, 77, 209–215, 2002.

Brown, R. E. & D. W. Macdonald. *Social Odors in Mammals*. New York: Clarendon Press, 1985.

Church, J. & H. Williams. "Another Sniffer Dog for the Clinic?" *Lancet*, 358, 930, 2001.

Cole, K. M. & A. Gawlinski. "Animal-Assisted Therapy: The Human-Animal Bond," *AACN Clinical Issues*, 11, 139–149, 2000.

Comparative Oncology Program of the National Cancer Institute. Website: http://ccr.cancer.gov/resources/cop/

Cooper, J. J., C. Ashton, S. Bishop, R. West, D. S. Mills, & R. J. Young. "Clever Hounds: Social Cognition in the Domestic Dog (*Canis familiaris*)." *Applied Animal Behavior Science*, 81, 229–224, 2003.

Coren, Stanley. *Why We Love the Dogs We Do*. New York: Free Press, 1998.

———. *How to Speak Dog: Mastering the Art of Dog-Human Communication*. New York: Free Press, 2000.

———. *The Pawprints of History: Dogs and the Course of Human Events*. New York: Free Press, 2002.

———. "Best Friends and Bed Partners." *Modern Dog*, 2 (1), 18–20, Spring 2003.

———. "Venus, Mars or Pluto?" *Modern Dog*, 2 (2), 30–33, Summer 2003.

———. "Laughing Dogs: Does Your Dog Enjoy a Good Joke?" *Modern Dog*, 2 (4), 26–30, Winter 2003.

———. *How Dogs Think: Understanding the Canine Mind*. New York: Free Press, 2004.

———. "Confidants to Kings: Consolers of Queens." *Modern Dog*, 4 (2), 59–63, Summer 2005.

———. "Medicine for the Mind: The Refugees of Hurricane Katrina." *Modern Dog*, 4 (4), 19–27, Winter 2005.

———. *The Intelligence of Dogs: Canine Consciousness and Capabilities*. New York: Free Press, 2006.

———. *Why Does My Dog Act That Way? A Complete Guide to Your Dog's Personality*. New York: Free Press, 2006.

———. "When a Marriage Goes to the Dogs, Who Gets Fido?" *Modern Dog*, 5 (3), 54–58, Fall 2006.

———. "Cloning Rover, Fluffy, and Snuppy." *Modern Dog*, 5 (1), 34–39, Summer 2006.

———. "The Universal Dog Translator." *Modern Dog*, 5 (4), 28–32, Winter 2006.

———. "Do You Look Like Your Dog?" *Modern Dog*, (6), 98–101, Fall 2007.

————. "Love Story: Can a Dog Really Love?" *Modern Dog*, 6 (1), 28–32, Spring 2007.

————. "Are There Dogs in Heaven?" *Modern Dog*, 6 (2), 32–36, Summer 2007.

Coren, Stanley & S. Hodgson. *"Understanding Your Dog for Dummies."* Hoboken, NJ: Wiley Publishing Inc., 2007.

Csányi, V. *If Dogs Could Talk,* New York: North Point Press, 2005.

Dale-Green, P. *Dog.* London: Rupert Hart-Davis, 1966.

Dawkins, R. *The Extended Phenotype.* Cambridge: Oxford University Press, 1999.

Derr, M. *A Dog's History of America.* New York: North Point Press, 2004.

Friedmann, E., A. H. Katcher, J. J. Lynch, & S. A. Thomas. "Animal Companions and One-Year Survival of Patients After Discharge from a Coronary Care Unit." *Public Health Reports*, 95, 307–312, 1980.

Glickman, L. T. et al. "Herbicide Exposure and the Risk of Transitional Cell Carcinoma of the Urinary Bladder in Scottish Terriers." *Journal of the American Veterinary Medical Association*, 224, 1290–1297, 2004.

Hare, B., M. Brown, C. Williamson, & M. Tomasello. "The Domestication of Social Cognition in Dogs." *Science*, 298, 1634–1636, 2002.

Hare, B. & R. Wrangham. "Integrating Two Evolutionary Models for the Study of Social Cognition." In: Bekoff, M. (Ed.); Colin Allen, (Ed.): *The Cognitive Animal: Empirical and Theoretical Perspectives on Animal Cognition.* Cambridge, MA: MIT Press. pp. 363–369, 2002.

Hart, B. L. & L. A. Hart. "Selecting Pet Dogs on the Basis of Cluster Analysis of Breed Behavior Profiles and Gender." *Journal of the American Veterinary Medicine Association*, 186, 1181–85, 1985.

————. *The Perfect Puppy.* New York: Freeman, 1988.

Katcher A. H. "Are Companion Animals Good for Your Health?" *Aging* (331–332):2–8, 1982.

Katcher A. H. & A. M. Beck. "Dialogue with Animals." *Transactions & Studies of the College of Physicians of Philadelphia*, 8, 105–112, 1986.

Krauss, H. *Zoonoses: Infectious Diseases Transmissible from Animals to Humans.* Washington, DC: ASM Press, 2003.

Leach, M. *God Had a Dog.* New Brunswick, NJ: Rutgers University Press, 1961.

Matthews, K. A. & B. B. Gump. "Chronic Work Stress and Marital Dissolution Increase Risk of Posttrial Mortality in Men from the Multiple Risk Factor Intervention Trial." *Archives of Internal Medicine*, 11, 309–315, 2002.

McCulloch, M. et al. "Diagnostic Accuracy of Canine Scent Detection in Early- and Late-Stage Lung and Breast Cancers." *Integrative Cancer Therapies*, 5, 30–39, 2006.

Mech, L. D. et al. *The Wolves of Denali*. Minneapolis: University of Minnesota Press, 1998.

Mery, F. *The Life, History and Magic of the Dog*. New York: Grosset and Dunlap, 1964.

Miklosi, A., E. Kubinyi, & J. Topal. "A Simple Reason for a Big Difference: Wolves Do Not Look Back at Humans, But Dogs Do." *Current Biology*, 13, 763–766, 2003.

Newby, J. *Animal Attraction*. Sydney: ABC Books, 1999.

Odendaal, J. S. "Animal-Assisted Therapy—Magic or Medicine?" *Journal of Psychosomatic Research*, 49, 275–280, 2000.

Page, J. *Dogs: A Natural History*. New York: Smithsonian Books, 2007.

Patmore, A. *Your Obedient Servant*. London: Hutchinson, 1984.

Pennisi, E. "How Did Cooperative Behavior Evolve?" *Science*, 309, 93–96, 2005.

Podberscek, A. L. & J. Serpell, "The English Cocker Spaniel: Preliminary Findings on Aggressive Behavior." *Applied Animal Behavior Science*, 47, 75–89, 1966.

Putney, W. W. *Always Faithful*. New York: Free Press, 2001.

Riddle, R. *Dogs through History*. Fairfax, VA: Denlinger's Publishers, 1987.

Ritchie, C. I. *The British Dog*. London: Robert Hale, 1981.

Rogers, K. M. *First Friend*. New York: St. Martin's Press, 2005.

Serpell, J. A. *The Domestic Dog: Its Evolution, Behavior and Interactions with People*. Cambridge, England: Cambridge University, 1995.

Stansfeld, S. A., R. Fuhrer, M. J. Shipley, & M. G. Marmot. "Psychological Distress as a Risk Factor for Coronary Heart Disease in the Whitehall II Study." *International Journal of Epidemiology*, 31, 248–255, 2002.

Stevens, J. R., F. A. Cushman, & M. D. Hauser. "Evolving the Psychological Mechanisms for Cooperation." *Annual Review of Ecology, Evolution, & Systematics*, 36, 499–518.

Thurston, M. E. *The Lost History of the Canine Race*. Kansas City: Andrews and McMeel, 1996.

Willis, Carolyn M. et al. "Olfactory Detection of Human Bladder Cancer by Dogs: Proof of Principle Study." *British Medical Journal*, 329, 712, 2004.

Wyatt, T. D. *Pheromones and Animal Behavior: Communication by Smell and Taste*. New York: Cambridge University Press, 2003.

Index

About the Author

STANLEY COREN, PH.D., F.R.S.C., is a professor of psychology at the University of British Columbia and a recognized expert on dog behavior and dog-human interactions. In addition to doing research and writing, he has appeared on many television shows, including *Oprah, Good Morning America, Dateline*, and *CBS Morning Show*, and is seen nationally in Canada as the host of *Good Dog!* He has been named Writer of the Year by the International Positive Dog Training Association and by the Animal Behavior Society. He lives in Vancouver, British Columbia, with his wife, Joan, three dogs, and her cat.